MW01491206

The Flowers of Good

The Flowers of Good

The Science and History of Marijuana Liberation

Sidarta Ribeiro

**Translated from the Portuguese
by Daniel Hahn**

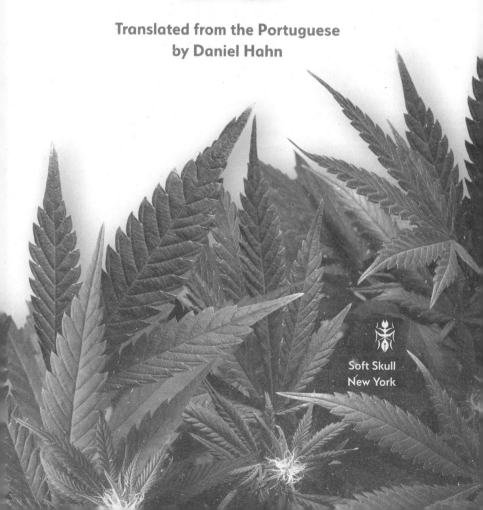

Soft Skull
New York

Copyright © 2026 by Sidarta Ribeiro
Translation copyright © 2026 by Daniel Hahn

First Soft Skull edition: 2026

ISBN: 978-1-59376-811-9

The Library of Congress Cataloging-in-Publication data is available.

Cover design and illustration by Victoria Maxfield
Book design by tracy danes

Soft Skull Press
New York, NY
www.softskull.com

Printed in the United States of America

10 9 8 7 6 5 4 3 2 1

Contents

Introduction

My brother, Júlio, and I first heard the word *"maconha"*—
marijuana—in the voice of our mother, Vera. Concerned as
she was about the dangers that came with preadolescence at
the start of the 1980s, she decided to broach the subject with
us preventively. *"Maconha* isn't for young people. Promise not
to use it. When you're older, if you really do want to try it,
we'll do it together, at home, but you're not going to do it out
somewhere with strangers. Agreed?"

"OK, Mom."

She had almost no experience with weed herself, but she
was open-minded and inspired great confidence. We were
still really young, twelve or thirteen, and we were receiving
an upbringing based on responsibility, freedom, and autonomy. We embarked on that simple agreement with total mutual trust, and it proved effective: We remained uninterested
in marijuana almost all the way through high school, while
some of our classmates were already smoking their first joints
in the undergrowth round the back of the school canteen.

As time passed, our interests began to diverge. My brother
liked mountain biking, I rode my bike on the track. He started

specializing in adventure and in risk, and I in books and role-playing games. However, even in this process of drifting apart, we still did some things for the pure pleasure of being together, like watching the TV series *Cosmos* on Sunday mornings, presented by the amazing astronomer Carl Sagan.

The family lull lasted until Júlio—seventeen now and hungry for the novelties of adult life—claimed our mother's promise. However, at that point, alarmed at the power of the rebellious youth that was starting to emerge, Mom faltered. She balked, she went quiet, she talked our ears off, she grumbled, and finally she went back on her word. The frank conversation stopped, and the train began slowly to come off the rails. Júlio started to drink with friends in bars and at parties, he took up cigarettes and became a fan of marijuana. There followed some experimenting with other substances, both legal and illegal, as well as two dangerous car accidents. Arguments at home were becoming ever more frequent, and after a few years and many conflicts, the family finally split apart.

My brother was kicked out of the house and I, like my stepfather, fervently supported my mother's decision. My sister, Luísa, then still a child, witnessed that split fearfully. Over the course of this painful process of distancing, it seemed perfectly evident that marijuana was to blame for our terrible family crisis. My deconstruction of that perspective is the story I want to tell in this book.

This deconstruction begins with a presentation of the therapeutic use of cannabis and its biological mechanisms, it moves through the history of the plant and the implacable persecution it has suffered, and it arrives at the economic, social, and political consequences of the gradual legalization of marijuana. The journey is threaded through with an autobiographical reflection on the role of cannabis in the construction of a better life, both for oneself and others. Enjoy!

The Flowers of Good

Marijuana Wins by *Ippon*

In Japanese martial arts, *ippon* is the decisive score that ends a fight and bestows the victory on whichever competitor was awarded it. In the fight over the therapeutic use of marijuana, *ippon* would come to be awarded to a very complex network of people including patients with epilepsy and their relatives, growers, scientists, health professionals, journalists, and politicians. After decades of clandestine rebellion, the defenders of marijuana saw their movement grow, become visible, and blossom.

Cannabis is a miracle of biological and cultural survival, a plant cultivated millennia ago on account of the exceptional textile fibers of its stems and the powerful resinous medicines of its inflorescences—which we'll call its flowers, for simplicity's sake. Those varieties rich in fibers but without the strongly psychoactive molecules are called hemp (*cânhamo*, we call it in Portuguese); while those rich in psychoactive resins are marijuana (and these in Portuguese were named *maconha*, an anagram of *cânhamo*). To make things easier, we will henceforth call both types of plant marijuana, unless there is some specific reason we need to differentiate between them.

In Europe, hemp was a naval product
that was essential for countries' defense.
(Chris Duvall)

In the sixteenth century, the clothes of European sailors and merchants were made of hemp, while the ointments of the medicine women and midwives of India and Africa were made of marijuana. From that time, hemp was the basis of almost all the canvases on which were painted every work of art framed on museum walls. In the eighteenth and nineteenth centuries, the plasters sometimes used on the backs of enslaved people to assuage the wounds produced by their foremen's whips were made of marijuana.[1] At the start of the twentieth, the bronchodilator cigarettes sold at pharmacies for treating asthma were made from marijuana. Moving ahead in time, on Brazil's initiative and under relentless pressure from the United States, marijuana was banned and excoriated as "the devil's weed." From the 1960s, however, despite all the persecution, consumption grew, until in 2022, in the United States, it overtook the consumption of tobacco.[2] Pushing back against all the stigmatizing of "potheads," a cannabis culture of peace and love spread right across the planet. On five continents, the most varied kinds of people now assemble at 4:20 to consume marijuana in an atmosphere of sharing, dialogue, and good humor. Today, medicines that have a basis in marijuana are increasingly exported by the United States, Canada, Portugal, and Uruguay, producing a great deal of health, employment, and income. Who'd have thought!

This incredible plant, so patiently constructed by the intelligence and tenacity of our ancestors, has weathered a global slander campaign that has lasted over a century. Despite all the persecution, however, cannabis and its main constituent molecules, called cannabinoids, are used today in the successful treatment—and with reduced side effects—of illnesses and disorders as varied as epilepsy, spasms, neuropathic pain, autism, cancer, depression, anxiety, Alzheimer's, Parkinson's, and Crohn's, among others. These applications relate to multiple metabolic and physiological effects of the molecules present in the plant—among them, effects that are analgesic, anti-inflammatory, antispasmodic, anti-ischemic, antiemetic, antibacterial, antidiabetic, and antipsoric and that stimulate the growth of bones.

We know today that the substances found in cannabis operate on our brains and immune systems through their resemblance to molecules produced in our own bodies. These small endogenous molecules, as well as the large receptor proteins located in the cell membranes, to which they are linked, collectively form the endocannabinoid system. So anyone who fears marijuana ought to consider that their own body is constantly producing a large quantity of molecules that closely resemble marijuana's. If you were to lose your endocannabinoid system, you would instantly lose immune responses and the capacity to feel hungry and feed yourself, to sleep, and to form memories. Marijuana produces effects in our bodies only because we synthesize substances that are functionally very similar.

Fortunately, the persecution of marijuana is ceasing to be acceptable in the twenty-first century. Its anti-epileptic effect, which has been described by science since the nineteenth century, was solemnly ignored by public opinion and by health professionals until about a decade ago; but when this finally changed, the plant was able to take its first step toward returning to medicine by the front door. Between 2000 and 2023, almost six times as many articles of biomedical research into cannabinoids were published as in the whole twentieth century. In the United States, the financing of research into cannabis grew from around $30 million in 2000 to over $143 million in 2018.[3]

Faced with these facts and with the ever-accelerating discoveries of marijuana's benefits, why do some people still insist on demonizing it? One of the worst problems of deliberate ignorance—an ignorance that clings stubbornly to prejudices—is that it tends to deepen with time, instead of lessening as new information is learned. Those who turn a blind eye toward novelties in science will tend to detach themselves more and more from reality and begin to inhabit a bubble of ideas that are increasingly outlandish. I have often come across interlocutors who are ill-equipped for the debate, because they simply haven't read or do not like anything positive that scientific research on marijuana has discovered. Actually, until the first decade of the twenty-first century, there was near unanimity in the medical field that marijuana and its derivatives ought not to become a part of the pharmacopeia, as better alternatives were already available on the market.

However, when the ignorance is involuntary and there is an intellectual honesty that is free from prejudice, it's never too late to rescue what has been left behind. One moving example is that of Dr. Sanjay Gupta, CNN's chief medical correspondent, who in 2013 released the first episode of a documentary series called *Weed*. His apology is worth reading:

> Long before I began this project, I had steadily reviewed the scientific literature on medical marijuana from the United States and thought it was fairly unimpressive. Reading these papers five years ago, it was hard to make a case for medicinal marijuana. I even wrote about this in a *TIME* magazine article, back in 2009, titled "Why I would Vote No on Pot."
>
> Well, I am here to apologize.
>
> I apologize because I didn't look hard enough, until now. I didn't look far enough. I didn't review papers from smaller labs in other countries doing some remarkable research, and I was too

dismissive of the loud chorus of legitimate patients whose symptoms improved on cannabis.

[. . .] I mistakenly believed the Drug Enforcement Agency listed marijuana as a schedule 1
substance* because of sound scientific proof.
Surely, they must have quality reasoning as to
why marijuana is in the category of the most
dangerous drugs that have "no accepted medicinal use and a high potential for abuse."

They didn't have the science to support that
claim, and I now know that when it comes to
marijuana neither of those things are true. It
doesn't have a high potential for abuse, and
there are very legitimate medical applications.
In fact, sometimes marijuana is the only thing
that works. Take the case of Charlotte Figi, who
I met in Colorado. She started having seizures
soon after birth. By age 3, she was having 300 a
week, despite being on seven different medications. Medical marijuana has calmed her brain,
limiting her seizures to 2 or 3 per month.

I have seen more patients like Charlotte first
hand, spent time with them and come to the realization that it is irresponsible not to provide the
best care we can as a medical community, care
that could involve marijuana.

---

* Schedule 1 drugs, according to the U.S. Drug Enforcement Administration, include drugs that are considered to have "no currently accepted
medical use and a high potential for abuse."

We have been terribly and systematically mis-
led for nearly 70 years in the United States, and
I apologize for my own role in that.[4]

Charlotte Figi (2006–2020) was an American girl with a
rare genetic disorder, severe myoclonic epilepsy of infancy,
or Dravet syndrome. This incurable illness causes progressive
motor and cognitive damage, which can lead to premature
death. Even when this outcome is avoided—many patients
with Dravet do reach adulthood—the behavioral and social
deficits tend to be dramatic, as the frequent interruption to
normal brain function caused by an epileptic seizure has an
amnesiac effect that severely hampers learning. In addition,
the excessive synchrony of neuronal activity that character-
izes an epileptic seizure releases an enormous amount of the
neurotransmitter glutamate, which in large quantities is toxic
and can end up killing neurons.

The need to keep Charlotte's epileptic seizures in check
led her doctors to prescribe high and regular doses of common anticonvulsant medications, which manage to reduce
the excess of synchronic neural activity by diminishing the
neurons' activity overall. While this strategy is effective for
mitigating seizures, it also creates a state of torpor that inhibits the child's normal development. Besides, the profound
depression of the nervous system caused by these drugs can
lead to a cardiopulmonary arrest, meaning that the relatives
of children with Dravet and other epilepsies need always to
have various kinds of resuscitation equipment at hand, a kind
of mobile ICU.

In families lacking the financial means, this situation is

a desperate one. And even those who are in a position to fund the necessary treatments are presented with a terrible dilemma: Don't treat the seizures and see the child wasting away in repeated spasms or keep them in a constantly sleepy state and running the risk of sudden death, under the effects of conventional medication. In both cases, the damage to development is huge, with significant emotional impact on everybody.

At five, Charlotte didn't attend school regularly, she moved about in a wheelchair, and she could barely speak. Her condition seemed only to be worsening when her mother, Paige, learned that the molecule called cannabidiol (CBD) might help. She visited the famous Stanley Brothers, producers of marijuana intended for the recreational market, and discovered that they had a variety of the plant that was high in CBD and low in tetrahydrocannabinol (THC), which had not been widely cultivated owing to its poor market value. Evocatively called "Hippie's Disappointment," this variety causes no alteration to the mental state because the THC that would produce such a thing is almost absent, while the CBD contained in the flowers inhibits it.

Treatment with the oil produced from this variety of marijuana changed Charlotte's life and her family's radically. The three hundred epileptic seizures she'd been having per week became three per month. Sleeping and feeding became more regular, and social interactions were now a possibility. Bit by bit, games became more frequent. Charlotte learned to ride a bike; she started attending school; she lived.

All this was possible because Colorado was—along with

Washington—a pioneering state in legalizing the recreational use of marijuana in the U.S., in 2012.[5] The legalization gave security to those cultivators who for many years had grown flowers in secret, and who were truly responsible for maintaining and disseminating information about the different strains of marijuana.

The "Hippie's Disappointment" strain was renamed "Charlotte's Web"—like the children's book—in honor of the girl who changed public perceptions of marijuana. When the news spread, families with similar diagnoses started to move to Colorado to benefit from CBD treatment. This emotional story finally traveled across the world in Sanjay Gupta's documentary series. Charlotte became an icon of the international movement for the therapeutic use of marijuana, bringing visibility to the similar dramas experienced by many other people with epilepsy. CNN would subsequently produce another six episodes of the *Weed* series, looking at a range of biomedical or cultural aspects of the plant.

When the corporate media and social media decided to disseminate information about the powerful anti-epileptic effects of cannabinoids, the tectonic plates of public opinion began to shift. How could anyone deny children with congenital epileptic syndromes the benefits of CBD, which was capable of inhibiting even hundreds of convulsive seizures per week? What justification could one give to a patient's mother or father for banning what was the best medicine for their daughter or son, even though it was something that could be grown at home?

The calls to lift the ban on marijuana for therapeutic

purposes began to swell, and soon public opinion started to wake up to the dangers of demonizing it. Between the first decade of the twenty-first century and 2019, the proportion of the U.S. adult population who supported the legalization of marijuana grew from 32 percent to 67 percent.[6] However, the opinion of supposed specialists remained quite conservative. I've lost count of the number of doctors—intelligent, well-informed people—who were seized by an acute lack of curiosity when it came to the subject of marijuana, perhaps once again fearful of public condemnation, or of being reprimanded within their close social circles. For many years, clinicians and researchers who were invited to share their opinions on the subject denied marijuana's therapeutic properties, while simultaneously exaggerating its risks to the lay public, stoking the moral panic.

By the time Charlotte died, in April 2020, at the age of thirteen, of pneumonia and a possible COVID-19 infection, hundreds of millions of people all over the world had already experienced the benefits of CBD for the treatment of conditions as diverse as epilepsy, chronic pain, anxiety, and insomnia. On learning of her death, Sanjay Gupta wrote: "Charlotte changed the world. She certainly changed my world and my mind. She opened my eyes to the possibility of cannabis being a legitimate medicine. She showed me that it worked to stop her crippling seizures, and that it was the only thing that worked." It is currently estimated that one in seven people in the United States use some CBD-based product, including cool drinks sold at gas stations. In Canada and Uruguay, marijuana was legalized for therapeutic and recreational uses and is commercialized or regulated by the state itself. These international experiences have actually increasingly called this schematic distinction between therapeutic and recreational uses into question. Or does finding pleasure in living not promote health, too?

Not that many years ago, undercover police officers patrolled the streets of New York ready to lock up anybody who dared to smoke a joint in public. Today, the city throngs with stores selling marijuana flowers and products of all kinds for adult use, including little colored candies made from molecules derived from the plant. The underground harvest of knowledge about marijuana exploded into a scientific revolution that shifted public opinion. The dogs barked, but the caravan moved on. Marijuana today is a commodity with an estimated global revenue of $59 billion, predicted to grow to $75 billion by 2029.[7] Things change . . . The struggle has ended, by *ippon*.

Brazil Is Still Lagging, but It's Making Progress

Brazil is going through a process similar to that already experienced in the United States, Canada, and Uruguay, albeit still remaining very out of step with the improvement in public access to the medication. Despite having been home to one of the world's greatest experts in therapeutic marijuana, Dr. Elisaldo Carlini (1930–2020), the country missed an opportunity to be a world leader in cannabis research, alongside Israel. But that is for the next chapter. Until the first decade of the twenty-first century, the debate seemed to have stagnated and was almost totally dominated by the most conservative strands of psychiatry. The stigma on marijuana was very great, and few people could see the need to research it. To make matters worse, there was a broad acceptance of various myths that induced moral panic, of the "marijuana kills neurons" variety.

I felt a considerable pressure myself to avoid the subject. In 2007, when Renato Malcher-Lopes and I co-authored a book for a somewhat specialist scientist readership about cannabis,[1] I chose not to disseminate it widely, for fear that I might lose research collaborators. However, in 2010 the musician Pedro Caetano, bass player in the band Ponto de Equilíbrio, was arrested for having planted marijuana for his own personal consumption.[2] His incarceration triggered the publication of a note from the Brazilian Society of Neurosciences and Behavior (SBNeC) in *Folha de S.Paulo* newspaper, signed by Cecília Hedin-Pereira, João Menezes, and Stevens Kastrup Rehen from the Federal University of Rio de Janeiro (UFRJ), and by me from the Federal University of Rio Grande do Norte (UFRN). Cecília was then vice president of the SBNeC, Stevens the treasurer, and I the organization's secretary. The note condemned the musician's being locked up and backed the calls for his release,[3] which occurred the following day.

The episode sparked an unprecedented scientific debate in Brazil around marijuana, which went well beyond lab benches and classrooms, both inside and out of the SBNeC. The fact that neurobiologists had joined the discussion reached major media outlets and threw the whole rhetorical claque of anti-marijuana prohibitionists up into the air.

In October 2010, *Folha de S.Paulo* programmed a heated debate between the psychiatrist Ronaldo Laranjeira, of the Federal University of São Paulo (Unifesp); Maria Lúcia Karam, former judge and member of the international organization Law Enforcement Against Prohibition (LEAP); the archaeologist Marcos Susskind from the NGO Amor Exigente (Love Demands); Renato Malcher-Lopes; and me. This encounter, moderated by the journalist Gilberto Dimenstein, was preceded by the publication of opposing articles in *Folha de S.Paulo* for and against marijuana legalization, with a right of reply, counter-reply, and counter-counter-reply. Tensions rose and increased the electricity around the scheduled duel.

I read and reread dozens of scientific papers to prepare for the debate, but when the discussion finally began, the arguments of the prohibitionists suggested they were firing blanks.[4] Renato and Maria Lúcia were dazzling. Biomedical science and legal wisdom had overcome the moral panic that had been created around marijuana, and the weed's detractors seemed unprepared for an audience who were not already converts to their beliefs. That is how I recall the event; doubtless the prohibitionists would think differently.

Released at around the same time, the documentary *Cortina de fumaça* (Curtain of Smoke), directed by Rodrigo Mac Niven, fired up the arguments about marijuana legalization in Brazil. With powerful testimony from patients, scientists, politicians, legal experts, and police officers, all of them anti-prohibitionists, the film recorded the moment when the activism that was organized at the Marijuana March caught alight.

This global event, occurring since 1999, had since 2008 seen the Brazilian street protests for cannabis legalization confronted with strong police repression, when marches convened in ten capitals by the Growroom collective, the largest Portuguese-language marijuana portal, and other growers' organizations were met with violence. This injustice and brutality only strengthened the movement, which spread right across the country like a fever of libertarianism and idealism. Just as happened with the LGBTQIAP+ pride parades around the world, each Marijuana March saw more and more people coming out of the closet, crossing barriers of race, class, and gender, to express their solidarity and fight for the sacred right to violate rules that are inhuman. Whether marching or not, directly or through metaphors, people as varied as Zé Celso, Fernando Gabeira, Eduardo Suplicy, Drauzio Varella, Julita Lemgruber, Rita Lee, Dilma Rousseff, Laerte, Geraldo Alckmin, Ailton Krenak, Luís Eduardo Soares, Marina Lima, Luciano Ducci, Otavio Frias Filho, Eduardo Giannetti, Paulo Teixeira, João Gordo, Marisa Monte, Marcelo D2, Mara Gabrilli, Andreas Kisser, Mano Brown, Patrícia Villela Marino, Luciana Boiteux, Renato Cinco, Jean Wyllys, Marielle Franco, Gregório Duvivier, Natália Bonavides, Sâmia Bomfim, Anitta, and Ludmilla took up the defense of weed.

The kernel for this great shifting of public opinion was solidarity with all those people who need marijuana for medical purposes and endure terrible pain because they are denied access to it. Patients like Juliana Paolinelli (1979) with neuropathic pain,[5] Thais Carvalho (1979) with ovarian cancer,[6] and Gilberto Castro (1973) with multiple sclerosis[7] all brought much visibility to the cause. Gilberto's words leave no room for doubt about its therapeutic value:

> Medicinal cannabis gave me back my life. It reduced the symptoms of MS without any side effects. [. . .]
>
> It clearly inhibits the evolution of the illness. This was one of the reasons I became an activist (one of Brazil's first), as that information needs to be passed on to save more lives. [. . .]

> For someone not expecting to live another
> five years, having made it through twenty thanks
> to marijuana is no small thing.

It's hard to convey just how revolutionary the meeting was between patients and their relatives who desperately needed marijuana and the maverick gardeners of cannabis, masters in the art of making it blossom. When these young people with dreadlocks and green fingers started to supply—for free—the raw material to cure so many kids, the pro-marijuana social movement came to a boil.

One of the first growers to sow this path was the lawyer Emílio Figueiredo (1978), who in 2010 responded to his father's appeal to donate flowers to a patient with multiple myeloma, who was unable to eat because of the intense nausea he was experiencing from chemotherapy. He was followed by other cases, and a powerful network of support and expertise began to take shape. Emílio learned to make marijuana tincture, oil, and butter for those who couldn't or didn't want to smoke. These experiences were shared with other growers and patients through the Marijuana Marches and communities like the Growroom collective. Its creator, William Lantelme Filho, recalled the spontaneous nature of that historic process:

> I never meant to be an activist. [It was] Grow-
> room that made me [become one]. In a way, my
> intention was already activism without my know-
> ing. I originally intended to plant a high-quality
> product, and taking care of the plant, knowing

all there was to know, seeing the whole of the plant's natural process. And I also always saw this possibility as a way of not needing to buy on the illegal market.[8]

As Emílio, William, and others of their generation were becoming exponents of cannabis activism in the country, a silent revolution in access to therapeutic marijuana was brewing. For the first time, the heroic resistance of the growers received the support of an influential part—albeit still a minority—of Brazilian society. White middle-class families were starting to create links of gratitude and political solidarity with young people across all social classes, but mostly with underprivileged students and skateboarders and motorcycle couriers, all united in their noncompliance and now protected by an unprecedented legal, scientific, and journalistic rearguard. It was this brilliant mixture of people and perspectives that discombobulated the public perception of marijuana's enemies.

In 2011, Brazil's Federal Supreme Court (STF) unanimously recognized the right to march to legalize marijuana, freeing users and sympathizers to express their consciences. For the first time, the experiences of patients and relatives began to be heard without being considered an apology for drug use. After that decision by the STF, the hardships experienced in Brazil by families of epilepsy patients gained visibility. Brazilian stories as moving and inspiring as that of Charlotte Figi in the U.S. began to appear in the country's leading papers and on mass-viewership programs on broadcast television.

The Clárian Case

We learned of a São Paulo girl born in 2003, who had Dravet syndrome, with autism, hypotonia, apathy, sleep apnea, repetitive behaviors, and long generalized convulsions. According to her mother, Cida Carvalho:

> A lot of children with Dravet don't make it to adolescence. Since Clárian was at risk of sudden death, Fábio and I were in a constant race against time—we even took turns sleeping. We needed to stretch out her trachea, and we were scared we'd lose her in her sleep. As for the hypotonia, the lack of sweating . . . until she was eleven, I had never seen her really sweat, which meant she wasn't regulating her body temperature which led in turn to more attacks of severe convulsions. I had to use bottles of water or damp towels to wet her head and neck, to avoid convulsions that could last more than an hour. Clárian's cognitive abilities were compromised: [She had trouble with] motor coordination, a lack of balance, her walking was impaired. She self-harmed, hit her head against the wall and tried to pull her teeth out by hand when she was upset.[9]

When the treatment began, when she was ten years old, Clárian Carvalho was unable to run, jump, climb stairs, or have a conversation. Because of her condition and the side effects of the conventional medication she was taking, the family routinely had to race to the emergency room, and they suffered from severe chronic stress.

"It was exhausting for all of us, we had no social life, we'd already tried various combinations of anti-convulsants with no success, and we knew that we might lose her at any moment," said her mother.

Clárian's situation was desperate and Cida didn't know what else to do, apart from throwing herself into intensive internet searches about treatments. In July 2013, she learned of the case of Charlotte and immediately asked her husband: "Shall we go find a pot dealer? If it works for her, I want to start planting marijuana!" Cida immersed herself in reading about the varieties of the plant and sent the articles she found to Clárian's neurologist, who hesitated at first, until she went to a congress in Boston, after which she agreed to endorse the treatment. Unlike other doctors, Dr. Maria Teresa Maluf Chamma was convinced: "My jaw dropped. I'm going to hoist this flag with you!"—and she encouraged that search for a cure. Cida reported:

> The doctor suggested that we "find other parents of children with Dravet, and set up an association." That was when I set up the first Facebook page, through which I would later have contact with doctors and other relatives of patients. Until I managed to get an international courier—Dr.

Maria Teresa herself. She was taking a vacation in Miami, I bought it and had it sent to the hotel where she was staying. And really bravely she brought it back for me on the plane, illegally. I got my holiday money from the bank where I worked to pay for the oil, which cost about R$2500 at the time. I was sure I wouldn't be able to keep it up, but we needed to know what effect this substance would have on my daughter . . . After a long battery of tests, Clárian took the first drops on 26 April 2014. Right away she spent eleven days without a single attack.

Well embedded now in a network of doctors, researchers, and supporters, Cida received an offer for the free supply of marijuana oil, from a secret network of growers in Rio de Janeiro who were sympathetic to the girl's plight. During the process of ensuring their daughter's treatment, Cida and Fábio went to Chile to visit the Mamá Cultiva association to prepare themselves technically for the extraction of the marijuana oil and to visit the Daya Foundation plantation. They also built bridges with science through the Brazilian Center of Information on Psychotropic Drugs (Cebrid), an organization that, under Professor Dr. Elisaldo Carlini, Brazil's leading researcher into therapeutic marijuana, organized a seminar in 2014 that brought a great deal of visibility to patients like Clárian. These links to the growers guaranteed her treatment. Between 2014 and 2017, Fábio traveled regularly from São Paulo to Rio de Janeiro to fetch the still-illegal medicine.

According to Cida:

After four months of using the oil, Clárian showed us her hands saying they were wet. I thought she'd been playing with water, but no, the palms of her hands and soles of her feet were sweating. That was the first time I saw my daughter actually sweat. From then on, she began to be able to regulate her body temperature and everything started to fall into place. After eight months of use, the improvement in her balance was very visible, Clárian was no longer leaning on people when she walked, with her knees half-bent, and she could go up and down stairs unaided. It was the first time she ever jumped without assistance on a trampoline at somebody's birthday. She could speak entire phrases that fit well into the family's conversation . . . Today, the convulsions have reduced by 80%. She has one or two attacks a month, lasting less than a minute.

The treatment also put an end to Clárian's sleep apnea, her muscle tone increased, and her cognitive abilities improved considerably. Her life flourished. Clárian became a very active girl and now, in 2023, she's learning to read and write.

In 2016, with the help of Rede Reforma (the Reform Network), a not-for-profit lawyers' collective comprising Ricardo Nemer, Emílio Figueiredo, Marcela Sanches, Cecília Galindo, and others, Clárian's family managed to secure a habeas corpus—which in this context simply means a waiver—to plant marijuana for therapeutic purposes. Soon afterward, they established Cultive: the Cannabis and Health Association. Together with Professor Elisaldo Carlini and Father Ticão, an important religious leader in São Paulo's East Zone, Cultive and a number of other organizations have since 2016 been running free training courses on the therapeutic uses of marijuana, an initiative that has already reached eighty thousand people.

Back in 2014, the documentary *Illegal: Life Won't Wait*, directed by Tarso Araújo and Rapha Erichsen, had introduced the general public to five families' struggles to treat their children with marijuana-based remedies. At the risk of being considered criminals, a couple called Katiele and Norberto Fischer managed to obtain CBD-based medicine to successfully treat their daughter Anny, then five years old. Anny presented with CDKL5 deficiency disorder, a rare genetic illness that causes severe epilepsy. In showing a white middle-class Brazilian family who were prepared to secretly transport the forbidden medicine on an international flight, the movie flung open all the contradictions of the drugs policy of the day. In Katiele's words:

> When we learned about CBD, we decided to import it. We were aware it was a product derived from *Cannabis sativa*, and consequently illegal in the country. But our despair at seeing our daughter having convulsions every day, at any moment, was so great that we decided to face up to things and bring it in, even if that made us drug-smugglers.[10]

Civil disobedience and loving devotion, combined in a single act of courage. While this whole struggle for the right to use marijuana for therapeutic purposes was taking place, scientists were simultaneously waging a number of battles just for the right even to study it. While scientific research into cannabis and cannabinoids has never been prohibited by the Drugs Law or by the international conventions to which the country was a signatory, it was difficult in practice to carry it out in Brazil, owing to the impossibility of producing the substance on domestic territory or importing it from the United States or Europe without very great trouble.

Still, this block was successfully challenged by a number of Brazilian researchers, beginning with the pioneering studies of the pharmacologist José Ribeiro-do-Valle and his then apprentice Elisaldo Carlini at the former Paulista Medical School, subsequently Unifesp.[11] Using the samples generously donated by Raphael Mechoulam and bravely transported to Brazil by Carlini, dozens of researchers were able to carry out their research. In the 1980s, cannabis research was established at the University of São Paulo in Ribeirão Preto, led by Antonio Zuardi, Francisco Guimarães, Jaime Hallak, José Crippa, and Alline Cristina de Campos. Over four decades, these researchers have focused principally on CBD, which, not being psychoactive, was easier to access and, consequently, to investigate.[12] They demonstrated a number of the molecule's therapeutic properties, eventually getting patents for a fluorinated synthetic version of CBD.[13]

Guimarães's testimony gives a sense of the challenges faced by the researchers:

I began my studies on CBD in the 1980s during my doctorate, as a parallel project suggested by my adviser, Dr. Antonio Waldo Zuardi. Following our first publication,[14] in 1990, I got a letter from Prof. Raphael Mechoulam praising the work and proposing a collaboration to test CBD derivatives. At that point it was already very clear in the scientific literature that CBD didn't produce the same effects as the main cannabinoid present in the *Cannabis sativa* plant, THC, including the production of dependency. However, the legislation was muddled and remiss on the subject, confusing [other] cannabinoids with THC. [. . .] We didn't yet have regulatory agencies like Anvisa. So for our collaboration, Prof. Mechoulam started sending us the compounds he produced (or, in the case of CBD, extracted from the plant), directly from Israel by mail. Those were pioneering years, in which the lack of more specific regulation meant studies being done in a gray area in terms of legislation (or the lack thereof).[15]

Over the course of the 1990s, other researchers did succeed in making some progress, notwithstanding these difficulties. In her doctorate at Unifesp's psychobiology department, Ester Nakamura-Palacios looked at the role of THC in the operational memory* of rats using a leftover batch of the

---

* Operational memory is transient and makes it possible to temporally put in order the execution of actions triggered by internal or external stimuli.

compound. Supervised by Jandira Masur and with Orlando
Bueno and Sérgio Tufik as her associate advisers, Ester
showed that the reduction in operational memory following
the administering of THC was reversible after the cannabinoid was withdrawn.[16]

In order to take the study further and investigate the
chronic effects of longer-term use, Ester submitted a research
proposal to the National Institute on Drug Abuse (NIDA),
the U.S.'s main organization supporting research into the effects of drugs, which historically had a strong anti-marijuana
bias. Surprisingly, her project was approved, and NIDA committed to supplying a quantity of THC that was carefully
calculated for the experiment.

However, the process that was required to get the substance into Brazil moved extremely slowly. By the time the
project was approved, in 1992, Ester had already transferred
to the Federal University of Espírito Santo (Ufes). This was
a problem because the substance needed to come into the
country via an institution with a history of relevant research into THC, and this was not yet the case for Ufes. So
a colleague of hers from her doctorate days assumed the
legal responsibility alongside Unifesp, which took charge
of the whole bureaucratic process. However, the THC only
reached Brazil in 1996, by which time Ester was leaving for
a postdoc in the United States. On her return, in 1998, one
of her first tasks was to fetch the THC from São Paulo in
order to get on with the project. That was when the scientist discovered that one of the four vials in her batch had
been handed over to another researcher without her knowledge. Her story makes very clear the enormous difficulty in

researching a substance that is banned, scarce, and therefore much fought over:

> And so I waited six long years to get the materials into my hands, only to find myself ultimately unable to carry out the original project, owing to the lack of a quarter of the amount calculated. I had to come up with a different project that required a smaller amount, and that would make it possible to look at short-term acute effects. That is what happened during Lívia Carla de Melo Rodrigues's doctorate, which she did under my guidance. In the years that followed, I got a lot of phone calls from people requesting samples of THC for their research. [. . .] Think it's hard to carry out research into THC? Imagine what it was like thirty years ago, having barely completed a doctorate, a woman, starting her career at a university with no tradition of research into psychotropic substances, especially THC. [. . .] I won't even tell you the whole saga (feeling like I've committed some crime) that went into transporting the shipment.[17]

Far from being exceptional, Ester's story is corroborated by other researchers in the field. Fabrício Pamplona, who did his master's and doctorate on cannabinoids under the guidance of Reinaldo Takahashi at the Federal University of Santa Catarina (UFSC), recalls the difficulty in obtaining the substances for his research:

> We never got authorization for anything, and
> when we tried, it went wrong. In reality, since
> they were synthetic molecules, that is, canna-
> binoids that are analogous to the phytocanna-
> binoids from plants, molecules with different
> names but similar effects to THC, the customs
> officers didn't actually know what they were and
> they passed without any trouble through cus-
> toms. Practically contraband [. . .].[18]

On the other hand, when the researcher tried to bring in a
drug called SR141716A, an antagonist of the CB1 receptor,*
which had the opposite effect to THC, the batch was appre-
hended and incinerated by customs officers because the word
"cannabinoid" appeared somewhere in the declaration at-
tached to the shipment by the pharmaceutical multinational
that had synthesized the compound.

Later, in the 2000s, Jorge Quillfeldt of the Federal University of Rio Grande do Sul (UFRGS) also managed to get
through the blockade using a clever strategy; instead of passing through the sluggishness of the system, he tried to dribble
around it.

> We started in 2002, which was still during the
> undergraduate years of Lucas Alvares—who is
> my colleague today—using agonists [activators]

---

* Technically, an inverse agonist.

like anandamide and synthetic inverse agonists
(i.e. blockers), like AM251, as tools for understanding the role of the endocannabinoid system
in different memory processes. Unlike agonists,
cannabinoid blockers were harmful to the phases
of consolidation and extinction and facilitated the
recall and reconsolidation of memories. These
studies did produce a good number of publica-
tions over the years, but the first article, in 2005,
took more than two years to be published. It was
turned down repeatedly for different reasons,
among them the simple fact that its findings
"didn't fit" the myth that "cannabinoids are bad
for you." In order to buy anandamide and other
cannabinoids, we simply gave only the substance's
complete chemical name—its commercial name
would nudge people's prejudices—and so we
never had any problems. Chemical names are
huge and abstract, beautiful—and correct.[19]

In spite of the courage, creativity, and resilience of all
these researchers, in the 2010s cannabis research in Brazil
was still gridlocked. The restrictions were many and there
was no clarity about when the decisive moment would come
that could break the impasse. There was a lot of jostling
about, involving the Ministry of Justice, the Ministry of Agriculture, and Anvisa. To complicate matters, the Federal
Medical Council (CFM) was totally opposed to the idea of
cannabis therapies, and the DEA in the U.S. blocked (and
still blocks) the sending of substances like THC to Brazil.

In 2015, when I was director of the Brain Institute of the Federal University of Rio Grande do Norte (UFRN), I joined forces with two neuroscientist colleagues to try to break through this blockage and understand how best to combine cannabinoids in order to treat convulsions. Claudio Queiroz, a talented researcher with a specialization in epilepsy and my colleague at the institute, had a lab involving different models of epilepsy in mice. Meanwhile, at Ufes, Professor Ester Nakamura-Palacios not only had experience with the pharmacology of cannabinoids but also had sufficient quantities of THC and CBD for us to begin our experiments.

Even though there was no guarantee that we'd be able to continue, we began our research one sunny Saturday. The first results were promising, but we soon realized that it wasn't simply a matter of comparing different proportions of THC and CBD. The most important comparison was to the complete extract of the marijuana flower, what in cannabis jargon is called "broad spectrum," that is, complex mixtures of cannabinoids, terpenes, and flavonoids just as they occur in the plant. Unfortunately, however, we saw no chance of getting access to these extracts, and our scientific curiosity seemed doomed to failure.

Then, all of a sudden, an opening appeared. In 2017, a young guy called Yogi sought me out in the hope of getting scientific support for requesting a preemptive habeas corpus that would allow him to plant marijuana for the treatment of his mother, who was suffering with Parkinson's. He told me that, after years of suffering harsh symptoms, his mother had been persuaded by him to try cannabis. In a matter of minutes after her first use, her life improved, as did that of the whole family. Cannabis then became her sole treatment, replacing a whole list of conventional medicines that carried all sorts of side effects. Now the family was requesting legal access to the plant.

I gave him the necessary help but then forgot all about the matter—I thought it unlikely that the request would be met. Just imagine my surprise when I arrived at work one day to find an envelope stamped by the Federal Court of Rio Grande do Norte notifying me of Judge Walter Nunes's decision. I almost fell out of my chair when I read these words:

> [. . .] that there should be granted, *initio litis* and *inaudita altera parte*, an order of safe-conduct in favor of the Patients to require that police officers of the State of Rio Grande do Norte refrain from any attempts to infringe upon their freedom of movement, due to the concurrent presence of requirements *periculum in mora* and *fumus boni iuris*, and also due to the necessity upon medical orders and recognized by the state, that the Patient MÁRCIA is in need of treatment with medicinal Cannabis, and also that they are prevented from seizing the seedlings of those plants used in the effecting of this therapeutic treatment. Given the above, I do HEREBY GRANT the required preliminary injunction, granting patients Márcia Maria Saldanha Pacheco and Yogi Pinto Pacheco Filho safe-conduct requiring that the enforcement authorities refrain from adopting any measures intended to restrict the patient's

freedom of movement, at the time of the impor-
tation of seeds, the production and cultivation
of the plant *Cannabis sativa* and *Cannabis indica*,
for exclusively medicinal purposes, sufficient for
the cultivating of 06 (six) plants, as well as the
transportation of the plants *in natura* between
the patients' home and the Brain Institute of the
Federal University of Rio Grande do Norte, for
parameterization with laboratory tests in order
to verify the quantity of cannabinoids present in
the plants grown, the quality and safe levels for
the use of their extracts.

The decision gave Márcia and Yogi the technical support
they needed to administer a cannabis therapy that was sci-
entifically grounded—as had been happening already, since
2016, with a number of families living in Rio de Janeiro, who
were part of the FarmaCannabis project run by Virgínia
Martins Carvalho, professor of toxicology at the Faculty of
Pharmacy at the Federal University of Rio de Janeiro (UFRJ),
in partnership with the Oswaldo Cruz Foundation (Fiocruz).
The habeas corpus also prompted the creation of the first
patients' association in Rio Grande do Norte—Reconstruir.

This decision, additionally, had a side effect of unblocking
our research into epileptic mice. Now we had access to different extracts of the plant and our scientific curiosity could
fly free. While the need for a cure shifted the tectonic plates
of cannabis therapies, it also opened up a legal pathway to
allow the need for greater knowledge to be dealt with. Led

by Claudio Queiroz, there followed years of experiments carried out by graduate students like Igor Praxedes, which finally showed the greater anti-epileptic effectiveness of the CBD-rich extract, compared to the THC-rich extract.[20] While the former reduced the duration and severity of attacks with every dosage that was assessed, the effects of the latter varied depending on the dosage, with an increase in attacks when on lower doses and a reduction when on higher ones. This pioneering research into the behavioral and electrophysiological effects of the broad-spectrum cannabinoid extracts was successful in comparison with pure CBD and diazepam, a conventional anticonvulsant.

In 2017, the Abrace Esperança Foundation in Paraíba became the first patients' association in Brazil to be granted authorization to plant, harvest, and process marijuana flowers to supply therapeutic oils to its members. Under the bold leadership of Cassiano Teixeira, Abrace today supplies marijuana oil to forty thousand patients.

The power of the mothers and fathers of patients is colossal, since they have the greatest moral authority to enact civil disobedience in favor of life. On the outskirts of Recife, in shacks without walls, the mothers and grandmothers of children with epilepsy cultivate the plant and make high-quality oil for their beloved kids. In defiance of the law or with the security of a habeas corpus, alone or in a group, people are planting justice with their own hands. The Reform Network estimates that the number of patients tended to by associations or individuals with habeas corpus exceeds 140,000.[21] To advance the democratization of access, in

2019 dozens of associations formed the National Federation for Therapeutic Cannabis (FACT), of which I am honorary president.

A lot of water has flowed under that bridge in the last ten years, in a fertile process of mobilizing civil society. Since 2016, a multidisciplinary group of activists has been granted access to federal research facilities through a link between Margarete Brito, from the Association to Support Research and Patients with Medicinal Cannabis; Eduardo Faveret, Ricardo Nemer, and Pedro Zarur from the Abracannabis Association; Cecília Hedin-Pereira from Fiocruz; and João Menezes and Virgínia Carvalho from UFRJ. At that time, Fiocruz was under the leadership of Paulo Gadelha, who set up a working group on medicinal cannabis led by Hayne Felipe, then the director of Farmanguinhos, Fiocruz's major medicine-producing unit. These institutions, in partnership, organized various events for scientific dissemination, as well as courses on growing and discussions about public health.

The neurobiologist Cecília Hedin-Pereira remembers the buzz when the scientists and growers met:

> One day we were invited by Pedro Zarur—me and Virgínia—to find out about what they were doing. We went to his house and learned that all of the Abracannabis growers donated flowers for making oil to the patients. Pedro Zarur had transformed the kitchen and living room of his house into a lab. They decarboxylated, they sieved [. . .] and they used the Magic Butter machine borrowed from Ricardo Ferreira, who

later brought one to Abracannabis. They had a whip-round to buy coconut oil and shared the final product. The Harle-tsu strain was brought by the orthopedist Ricardo Ferreira from California and he gave it to Pedro and a friend for them to germinate and select which was the hardiest strain. They made three hundred clones they gave to Cidinha [Carvalho], and the rest went to a number of growers around Rio de Janeiro and other states. Margarete [Brito] among them.[22]

Marijuana is like the genie that has escaped the lamp: Once freed, it can't ever be contained again.

In 2019, the respected doctor Drauzio Varella—the Brazilian equivalent of Dr. Sanjay Gupta—posted on the internet (at 4:20 in the afternoon) a series of five YouTube episodes discussing the therapeutic use of marijuana, which amassed more than 7.5 million views over the next four years.[23] In 2023, Rapha Erichsen launched her second documentary about therapeutic marijuana, *Sofia's Other World*, depicting the struggle of another family who had appeared briefly in *Illegal*: Margarete Brito, Marcos Langenbach, and their daughter Sofia, who has the same syndrome as Anny.

Beginning with the donation—by growers from the Abracannabis association—of plant seedlings with high CBD content, in 2016 Margarete and Marcos had obtained the first habeas corpus in Brazil for the planting of marijuana, to use for treating Sofia. Gradually they began to increase their production of marijuana oil, donating the surplus to other patients. Along with the pediatric neurologist Eduardo

Faveret, who was then running the Epilepsy Center at the State Brain Institute, they mobilized an extensive network of relatives of children with epilepsy and founded Apepi. In 2020, this association was granted a legal authorization, in a case against Anvisa and the federal government, to cultivate, transport, handle, research, and supply marijuana oil to families across the country. In August of 2023, as I write these words, Apepi has around eight thousand members and takes on more than six hundred new ones each month.[24]

In the context of the regulatory vacuum and government inaction over the therapeutic use of marijuana in Brazil, Abrace, Cultive, and Apepi are just a few examples of the real explosion of patients' associations in every part of the country. Dozens of other smaller groupings, often organized around a single family with considerable inclination for the fight, have contributed to saving lives, while the public authorities still fail to meet their responsibility to provide the necessary medication through the country's Unified Health System (SUS). In 2023, however, various initiatives to include cannabis therapies in SUS were launched by legislative assemblies in several states of the country.

   In 2022, the sixth subpanel of the High Court of Justice (STJ) gave unanimous authorization for three people to grow marijuana for therapeutic purposes in their homes. Since then, that decision has served as guidance for other legal decisions all over the country, leading to a substantial increase in the granting of habeas corpus for domestic planting. In spite of the success in achieving legalization, this strategy is limited, as it doesn't deal with the majority of people who cannot, or don't want to, or don't know how to attain that legalization.

When a law granting exceptional permission to a few people is starting to be generalized, it's time to ensure it for all. Bill no. 399/2015, passed by a special committee of the Chamber of Deputies by a margin of just one vote and still waiting to be voted on in the plenary, proposes objective conditions for the growing of marijuana for therapeutic purposes on domestic soil, whether by companies or by patients' associations, as well as authorizing the public health care system to grow and harvest marijuana for the creation of phytotherapeutic products.* While this bill does mark an important step forward, it favors large corporations by stipulating conditions that would exclude most of the associations with less economic power. In addition, it does not consider the right to domestic growing.

Still in the field of possible advances, in December 2022, Anvisa unanimously authorized the Brain Institute at UFRN—from which, by this point, I had stepped down as director—to import, store, and germinate marijuana seeds, as well as to grow them and process them for research on cannabis therapies for refractory epilepsies using animal models. This decision made UFRN the first institution in the country to break through the block against research into marijuana without resorting to any subterfuge. This was possible because the university's chancellor, José Daniel Diniz de Melo, acted cleverly with Anvisa to assert the right to research that is enshrined in the legislation. In his words, that approval

---

* At the time of writing this book, the project had not yet been concluded.

"represents an important step in advancing the research developed at UFRN and a historic frontier for Brazilian science."[25] According to the project's coordinator, Claudio Queiroz, with whom I shared the experiment begun in 2015,

> UFRN can begin a robust program of research into cannabis for therapeutic purposes in a pre-clinical phase, contributing to scientific, social and economic development [. . .].
>
> The plan is to evaluate the existence of synergic effects among phytocannabinoids in the control of neuronal excitability [. . .].
>
> Thanks to Anvisa's historic decision, Brazilian society can see that it's possible to develop cannabis research, on domestic soil, with safety, traceability and responsibility.[26]

In parallel with the pro-cannabis movement in the public sector, there was significant growth in private interest in cannabis biomedicine. Researchers and entrepreneurs like Patrícia Villela Marino (The Green Hub), Claudio Lottenberg and Dirceu Barbano (Zion MedPharma), José Roberto Machado (OnixCann), Viviane Sedola (Dr. Cannabis/Cannect), and Bruno Soares (Ease Labs) have invested tens of millions of reais into the pharmaceutical industry and services connected with cannabinoid therapies. Like everything in life, this has a good side and a bad side. The good side is the power of cash to build amazing things, as the composer and singer Caetano Veloso would put it. The bad side is that many of the large-scale investors are increasingly removed from the social movements of patients and their associations, that is, from those who actually forged the path toward a broad acceptance of cannabis therapies. Even though they are beneficiaries of the civil disobedience of those who struggle at the grassroots for the right to health, the marijuana capitalists— with honorable exceptions—tend to keep their distance from these pioneers, arguing that they are illegal organizations that don't obey the rules of "corporate compliance."

Taking a step back to consider society as a whole, it's obvious, in spite of all the difficulties, that the (strategic or merely tactical) alignment of the legitimate interests of patients and relatives, growers, scientists, and business is defeating—by *ippon*—the ban on therapeutic marijuana in Brazil. We're still just at the start of a profound change. With comprehensive and multifaceted impact that ranges from neurology and psychiatry to oncology, endocrinology, and geriatrics, marijuana is to twenty-first-century medicine what antibiotics were for twentieth-century medicine.

If you think that claim is exaggerated, it's worth reading the statement made in 1997 to the U.S. Congress by Lester Grinspoon, professor of psychiatry at Harvard:

> Mr. Chairman and members of the subcommittee, I appreciate the opportunity to appear before you this morning and share my views on the use of marijuana as a medicine.
>
> [. . . In] 1928, Alexander Fleming discovered penicillin. This discovery was left on the shelf until 1941 when the pressures of World War II and the need for another antibiotic besides sulfonamide

compelled two investigators to look at it, and in
just six patients they demonstrated how useful it
was as an antibiotic. In fact, penicillin went on
to earn a reputation as the wonder drug of the
forties.

Why was it called a wonder drug? One, because it was remarkably non-toxic; two, because
once it was produced in large quantities, it was
very inexpensive, and three, because it was remarkably versatile; it would treat everything
from pneumonia to syphilis. Cannabis bears
some remarkable parallels with penicillin. First
of all, cannabis is remarkably safe. Although it is
not harmless, it is surely less toxic than most of
the conventional medicines it could replace if it
were legally available. Despite its use by millions
of people over thousands of years, it has never
caused an overdose death.

Secondly, cannabis, once it is free of the prohibition tariff, will be quite inexpensive. [. . .] And
then like penicillin, it is remarkably versatile. It is
useful in a number of symptoms and syndromes.[27]

Nearly three decades after this forceful statement, the fact
that the subject is still hard to discuss in Brazil demonstrates
the huge scale of the country's delay in properly regulating the therapeutic use of marijuana. If you are still unsure
what your position should be, just consider what would have
happened had penicillin not been widely taken up during
the Second World War . . .

The Flower of the Ganges
Was Born in China

But where did this plant, which is so very useful, actually come from? Marijuana is not a natural gift nor a divine one, but rather a product of the close and persistent relationship between the cultivated plants and the cultivating humans. Through the whole Paleolithic period, our ancestors developed tools for various functions out of stones, sticks, skins, and bones. And then, between thirty and twenty thousand years ago, when the final Ice Age was still lashing most of the human populations in Eurasia with hailstones, some of our forefathers had a brilliant idea that would transform our society forever: to use another living thing as a tool, transforming a dreadful enemy into a bosom friend. Great foresight must have been needed to look at the son of a wolf and see a faithful guard dog, capable of protecting people rather than attacking them. Through artificial selection and social integration, those ravenous wolves that were always lying in wait for the elderly and children around the outer edges of the camps were genetically transformed and psychologically co-opted to protect—with their own lives, if necessary—the physical integrity of the most vulnerable members of a human family. Our ancestors made one of the most dangerous predators into a versatile ally, capable of performing countless roles that were fundamental to human life: shepherd, guard, soldier, draft animal, tracker, guide, friend, therapist, court jester, pillow, rescuer, and supplier of alcoholic beverages to those trapped under alpine avalanches.

It's worth emphasizing that this was not a chance discovery of the 1,001 uses of preexisting species—the different species of wolf—but rather the progressive transformation of its genome[*] through the selective crossing of those organisms with a particular desirable phenotype, such as the presence or absence of aggressiveness, the abundance or lack of fur, etc. Once this selection had been carried out by the conscious action of human beings, intentionally and transgenerationally, we saw the creation—or rather, we should say, invention—of hundreds of different genotypes,[†] corresponding to hundreds more different phenotypes.[‡] Our close relationship with dogs forged an unprecedented alliance between intelligent and social species, which was highly beneficial to both. It also created a different way for us to look at other living creatures, not only as people who desire or fear different organisms, but as people who build them into quite familiar life forms. Since the invention of dogs, we have never been the same.

---

[*] The genome is the hereditary collection of genetic information of a species, stored in the structure of the deoxyribonucleic acid (DNA).

[†] The genotype corresponds to the complete set of an organism's genes.

[‡] The phenotype is the group of features observable in an organism, which might be behavioral, morphological, physiological, and biochemical in nature, resulting from the interaction of the genotype with the environment.

With the end of the last Ice Age, around twelve thousand years ago, new geographical frontiers opened up for human occupation, and a torrent of new domestications followed. The domestication of animals that had been positioned ecologically as predators—wolves—was followed by that of animals positioned ecologically as prey. Our relatives then began selectively to cross the ancestors of what we now call sheep, goats, pigs, cows, llamas, horses, camels, chickens, ducks, and many other animals. Parallel to this, between ten thousand and three thousand years ago, we domesticated almost all the plants that make up our diets now: wheat, peas, olives, rice, sugarcane, bananas, sesame, eggplant, figs, oats, sorghum, potatoes, corn, beans, pumpkins, palm oil, sunflowers, coffee, and almost everything else that humans still eat today.

It is striking that, for each one of the domesticated species, which include animals and vegetables, but also fungi and bacteria, there arose a huge number of strains, with specific flavors, functions, and effects. After all, it makes a lot of difference whether you're being protected by a Rottweiler or a pinscher.

It was in this context of genetic and cultural ultra-specialization, in the service of humans' many needs, that cannabis appeared from out of a genetic stock that is currently represented by wild plants and strains that are native to China. The word "cannabis" seems to come from the Hebrew/Aramaic expression *"kaneh bosm,"* which can be translated as "aromatic cane." *Cannabis sativa*, then, means "cultivated aromatic cane," in Latin. Therefore, the Neolithic domestication of the wild species of cannabis that would give rise to the hundreds, possibly thousands of marijuana strains grown on the planet today has a real parallel with the domestication of wolves into more than 350 breeds of dog—hence the bit of wordplay that I use, when I refer to the plant as "canine-nabis."

The comparative analysis of marijuana genomes that originate in various different regions of the planet has shown that it was domesticated for the first time around twelve thousand years ago, in eastern China.[1] This dating at the very beginning of the Neolithic period makes marijuana a good candidate for the oldest crop in the world. Excellent for the production of rope, fabrics, paper, food, and medicines, marijuana coevolved with the human species over the long period of creation and development of grazing and agriculture, until it became an undisputed love affair during the Bronze Age (3300 to 1200 B.C.E.). In time, the plant was spread beyond the Himalayan mountain range, reaching India, Afghanistan, and Siberia.

At the end of the Bronze Age, we begin to see a genetic separation of the two main types of cannabis: hemp and marijuana. Research has identified various genes responsible for this divergence, such as those controlling branching patterns and the biosynthesis of cellulose, of lignin, and of psychoactive cannabinoids. This process produced three subspecies: *Cannabis sativa*, *Cannabis indica*, and *Cannabis ruderalis*—which would have a great impact on those regions' populations.

One of the traces of this impact is the story of the Ice Maiden, a mummy found in the frozen Altai mountains in Siberia, in an incredibly rich ritual burial that can be dated to around 500 B.C.E., since the conditions of the tomb were preserved beneath the permafrost. A reconstruction of the mummified body suggested that she was a beautiful woman

The preparation of bhang in Turkmenistan (c. 1871, by
K. P. Von Kaufman, Turkestansk al'bom [1871–1872].
Courtesy of the Library of Congress)

Hmong people weaving hemp on a loom (Chris Duvall)

aged around twenty-five. Her skin was covered in elaborate tattoos of fantastical creatures, she wore a huge headdress and precious clothing in colored silks, and she was surrounded by three horses, probably sacrificed to accompany her on the journey beyond life. The rich possessions placed in the tomb and the sophisticated tattoos suggest a person of high social status, possibly a priestess or political leader.

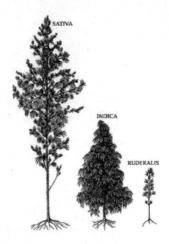

Comparison of *Cannabis sativa, indica*, and *ruderalis*
(Wikimedia Commons)

An analysis of the mummy using magnetic resonance imaging detected signs of breast cancer, with probable metastases in the thoracic vertebrae. Curiously, among the various precious objects found in the tomb, such as jewels, gold, and a Chinese mirror, there was a receptacle holding something that might have been marijuana. Could the Ice Maiden have used the plant to treat the terrible pains, anxiety, insomnia, and loss of appetite that come with the growth and spread of cancer? Might she have tried to use weed to stop the tumor growing? Recent research would suggest so.

Cannabis sativa (Wikimedia Commons)

This ancient use of cannabinoids is related to increased appetite and weight gain in oncological patients.[2] Cannabinoid receptors and the endogenous molecules they connect to are produced to excess in various tumor tissues.[3] In addition, the increased quantity of endocannabinoids is frequently associated with the aggressiveness of a cancer.[4] Various cannabinoids present in marijuana show anticarcinogenic activity in *in vitro* or *in vivo* models for skin cancer, prostate cancer,

breast cancer, and gliomas, among others.[5] The anticarcinogenic activity of phytocannabinoids is linked to their ability
to regulate signal pathways that are critical for the growth
and survival of cells, leading to an inhibiting of the cell's
proliferation and migration, to an inhibiting of angiogenesis[*]
and metastasis,[†] and to the inducing of apoptosis.[‡] Two clinical trials on the use of CBD to treat glioblastoma report very
promising effects in terms of the regression of this most ag-
gressive type of cancer.[6] Another clinical trial on 119 patients
with different solid tumors showed a reduction in tumor size
in 92 percent of patients when CBD oil was administered.[7]

Ah, science—so powerful and yet so slow! The cannabis
revolution is an incredible rescuing of the past. In China, there
is archaeological evidence of marijuana being grown for the
fibers used to produce fabrics, rope, and paper six thousand
years ago. These materials were found, for example, in the
tomb of Emperor Wu of the Han dynasty (156–87 B.C.E.)[8].
The ideogram for marijuana, ma (麻), shows two plants under a protective covering. Magu, the Taoist goddess of food,
cure, and female protection, who was (and still is) worshipped
in China, Korea, and Japan, is associated with hemp. The
therapeutic use of the resins of marijuana was also developed
early in China, and is included in the Bencaojing, the oldest

---

[*] Angiogenesis is the process of formation of new blood vessels from other
previously existing ones.

[†] Metastasis is the process by which cancer cells detach from the main
tumor, enter the bloodstream or the lymphatic system, and spread to
other tissues or organs.

[‡] Apoptosis, or programmed cell death, is a self-organized process that
plays the essential role of eliminating superfluous or defective cells.

pharmacopeia in the world, compiled around two thousand years ago and attributed to the emperor Shennong, the mythical creator of Chinese agriculture and medicine, who is said to have lived 4,700 years ago.[9]

LEFT: *The Immortal Magu with Deer and Peach Tree.* Silk tapestry, China, Ming or Qing dynasty, c. 1575–1725 (© Freer Gallery of Art, Smithsonian Institution). RIGHT: Seshat, the Egyptian goddess of writing and knowledge (Wikimedia Commons)

In this compilation of very ancient oral traditions, marijuana is referenced for use in treating rheumatic pains, constipation, and gynecological disorders. The founder of Chinese surgery, Hua Tuo (who flourished from the late second century C.E. to the early third century C.E.), prescribed marijuana as an anesthetic for painful operations.[10] We know today that marijuana does indeed help in all these situations, for its analgesic and anti-inflammatory properties.

But in spite of its Chinese origins, it was in India that the use of marijuana as a medicine for body and soul took root most deeply, around three thousand years ago. The plant is an integral part of medicinal treatments and religious rituals to the north and south of the Ganges, the river that gave marijuana one of its main nicknames, coined in India and used particularly in Jamaica, in reggae culture: ganja.[11]

Marijuana plays a central role in Indian medicine, mythology, philosophy, and meditative practices. Over the last three millennia, it has been used in India almost as a panacea, recommended for the treatment of pains, epilepsy, tetanus, rabies, anxiety, rheumatism, infections, parasitic infections, diarrhea, colic, loss of appetite, and asthma. The Atharvaveda, a collection of sacred texts dated between the twelfth and ninth centuries B.C.E., lists marijuana as one of the five sacred plants, capable of promoting liberty and joy. You can also find ancient references to the therapeutic use of marijuana in the Satapatha Brahmana (between the eighth and sixth centuries B.C.E.) and Sushruta Samhita (between the fourth and sixth centuries C.E.).[12] Vangasena's "Compendium of the Essence of Medicine," an Ayurvedic text from the eleventh century, prescribes cannabis for increasing happiness and a long healthy life.[13]

Another place where the therapeutic use of marijuana seems to have been established early is in Egypt.[14] The Ebers Papyrus, dated to around 1500 B.C.E., contains a prescription for marijuana as a vaginal anti-inflammatory. Traces of marijuana were found in ancient Egyptian mummies,[15] and pollen from the plant was detected in the tomb of Ramesses II, who died in 1213 B.C.E.

Seshat, the "Mistress of the House of Books," the ancient Egyptian goddess of writing and of knowledge, associated with expertise in areas as diverse as accounting, astrology, and architecture, was typically represented as a scribe with one of her arms extended in the act of writing, while over her head lies a leaf that looks very much like a marijuana leaf, though other interpretations are possible.

Like the Egyptians, the Assyrians, Persians, Dacians, Scythians, and Hebrews all used marijuana as an intoxicating incense. Excavations carried out in Tel Arad, an Israeli archaeological site close to the southern border of the old Kingdom of Judah, discovered traces of marijuana in a 2,700-year-old sanctuary. The discovery supports the notion, which is argued by some historians, that marijuana was an integral part of the religious rituals practiced in ancient Judaism.

Within Islam, the Sufi tradition celebrates marijuana consumption as a form of inducing their spiritual trance. Some theologians even go so far as to attribute the discovery of marijuana to Sheikh Haydar, an important Sufi leader from the twelfth century. Since then, the preparation known as hashish, basically a concentrate of resins that come from the female flowers of marijuana, has spread from the Muslim world. In spite of the relationship of Sufism with the spiritual and psychoactive use of the flowers, hegemonic currents in Islam worshipped only the plant's fibers. From the twelfth century, the use of hemp for producing paper was a hallmark of Islamic culture, which spread along the shores of the Mediterranean and around the Sahara in the caravans of Arab merchants.

After making its way to Iran, the Balkans, Egypt, Ethiopia, sub-Saharan Africa, and southern Europe, marijuana finally arrived in the Americas in the sixteenth century, in the hands of both Europeans and Africans.[16] In the centuries that followed, the colonizers and their descendants spread the growth of hemp, for textile purposes, across America. Thomas Jefferson stated that "hemp is of first necessity to the wealth and protection of the country." George Washington, the country's first president, apparently said—though the attribution is contested—"Make the most you can of the Indian hemp seed and sow it everywhere."[17] In order to supply its naval industry, the Portuguese crown in 1783 created, in Rio Grande do Sul, the Real Feitoria do Linho Cânhamo (the Royal Factory of Hemp Linen).[18]

While the whites were having hemp grown for the production of clothing, sacks, ropes, canvas, and sails for ships, the enslaved Blacks were arriving from Africa carrying the seeds of plants rich in resins that were highly valuable for soothing the immense physical, emotional, and spiritual trauma caused by that diaspora.[19] The first great wave of Africans brought by force to Brazil at the start of colonization came from Angola,[20] which explains why the Portuguese words "*maconha*," "*liamba*," and "*diamba*" that are popularly used in Brazil to refer to cannabis are Angolan in origin.* It

---

* The Kimbundu word "*mariamba*" means "cannabis" and is directly related to the words "*marimba*" in Colombia and Cuba, "*marijuana*" in Mexico, and "*maconha*" in Brazil.

also explains why the use of smoked marijuana was banned by Rio de Janeiro's Municipal Chamber in 1830.

Despite this prohibition, until the 1930s, the use of marijuana as a sacred plant in religious practices of Afro-Indigenous origin, such as Candomblé and Catimbó, were relatively common in Brazil.[21] Outside the *terreiros* (the religious "houses"), usually on Saturdays, Black men would gather among their elders to smoke marijuana in "assemblies" or "confraternities" in which they reviewed the week, mixing amusement, trance, and the discussion of community issues.

This old Afro-Brazilian custom resembles the "reasoning sessions" in the Rastafari religion in Jamaica, in which marijuana plays an important role within the discussion of communal problems but also in the chants and prayers that can encourage a person's greater closeness to the god Jah.[22] For the followers of Rastafarianism, marijuana is a sacrament attributed to Ethiopian traditions as old as Jesus Christ, capable of promoting the bond of divine love that unites all people through the expression "I and I," which suggests the internal dialogue with one's own consciousness, but also "you and I," "us," in the external dialogue with other consciousnesses.

Even while it was being tirelessly persecuted, marijuana spread across the Americas and today forms part of the habits of peoples as diverse as Mexicans, Canadians, and U.S. citizens. In the deepest heart of the Amazon jungle, marijuana is still hated as an illegal drug but also loved as a sacred plant with countless uses for humans, having even been incorporated into the cosmologies and traditions of Indigenous people that are full of life force, such as the proud Tenetehara, the brave Krahô, and the magnificent Huni Kuin.[23]

None of this should be surprising. Three thousand years ago, at the end of the Bronze Age, our ancestors were already domesticating specific varieties of marijuana, which were cultivated separately for the production of medicines or fibers.[24] Today it is estimated that thousands of genetic varieties of these plants exist, differentiated by their chemical compositions. We are experiencing the African principle of *sankofa*: retrieving what has been left behind, possibly creating an ancestral future that could even postpone the end of the world, in the wise words of Ailton Krenak.[25]

Liamba...

psychic drug of the jungle

This poster, depicting *liamba*, "psychic drug of the jungle," reveals a certain stigmatizing of cannabis. (Chris Duvall)

In the debate over the legalization of the therapeutic use of marijuana, it's common to hear the mistaken analogy to jararaca pit vipers to justify the ban on growing the plant. To understand the argument, you need to keep in mind that the jararaca's venom contains a peptide that reduces the blood pressure. A synthetic analog of this peptide, discovered in 1965 by the Brazilian scientist Sérgio Henrique Ferreira, is available in drugstores as a hypertension medicine. The terrible analogy used by the prohibitionists is that nobody should ever have a marijuana plant in their home, just as nobody should have a jararaca, because both the plant and the reptile are extremely dangerous.

This unsound reasoning is based on the fallacy that just chemically extracting a single pure active principle would be all that's needed to give patients therapeutic safety. According to this line of thought, the jararaca is a very dangerous animal, but a single molecule isolated from its venom— an inhibitor of the angiotensin-converting enzyme called captopril—is beneficial. And so by analogy, marijuana is a dangerous plant, but only one single type of molecule analogous to that isolated from its flowers—CBD—would be medically useful.

It's not hard to see the intellectual falsehood: Unlike the venom of the jararaca, the molecules of marijuana are harmless to almost the entire adult population, with the exception of people in particular risk groups (like any substance, it does have its risk groups). The biochemical interactions between the components of jararaca venom act toward death, while the interactions between the components of marijuana act toward life. With all due respect to jararacas, I prefer the company of marijuana.

There is a variant of this flawed argument that considers marijuana as analogous to radioactive isotopes: yes, useful to medicine, but so dangerous that they require strict controls, with every plant and every flower constantly monitored,

everything duly labeled with QR codes, multiple barriers
to access, video cameras, and armed security. All this fuss
around a simple, useful plant that has been cultivated since
Neolithic times is a sign of ignorance and moral panic.

Marijuana is not the deadly venom of a jararaca, nor is
it radioactive cesium-137. An alternative metaphor, which
is much more moderate and useful to the discussion that we
need to deepen, is that marijuana is to plants what dogs are
to animals: They are human inventions, created to satisfy human needs. This analogy is not merely pedagogically useful,
it's also rigorous in scientific terms. As we saw at the start
of the chapter, our relationship to marijuana occurs within
the broad context of humans' domestication of animals
and plants. Would it make any sense to ask whether dogs
are beneficial, whether we ought to consider legalizing them?
There's something tragicomic about the whole thing.

Of course, we know that our contact with dogs and with
marijuana can go wrong, if wrongly conducted. Our inventions are not inherently good, since their effects depend on
what we do with them. Despite their having been genetically
selected to solve human problems, the countless types of dog
and of marijuana can have distorted uses. Protection of people who are attacked by dogs or whose use of cannabis is problematic will not come from an attempt to ban these organisms
that humans have shaped but from a scientific familiarity with
potential harms, risk groups, and protective measures. Another domesticated species might help to explain the issue.
Some people are allergic to or intolerant of gluten. Should
we therefore ban the planting and consumption of wheat?
Banning intolerantly does a disservice to all.

The indisputable fact is that marijuana is an extremely useful plant, which we have inherited from those who were here long before us. Failing to honor, celebrate, and disseminate this ancestral legacy is simultaneously heretical and stupid. The real problem, actually, is not simply about the legalization of marijuana. For instance, its therapeutic use has been effectively authorized in Brazil since 2017, when a cannabis-based nasal spray began to be imported and sold in pharmacies to treat the spasticity prompted by multiple sclerosis. A 30 ml vial containing 27 mg/ml THC and 25 mg/ml CBD is currently sold in Brazil for the equivalent of around US$480. Marijuana has therefore been perfectly legal in Brazil for years, but only to those who are materially better off.

The real problems are to do with the lack of access and with social stigma. Which is why more and more people are petitioning to ensure that SUS import and supply exceptionally expensive marijuana-based medications, despite the significantly reduced cost of producing the broad-spectrum oil domestically, which remains illegal. For Brazil, importing marijuana is as unnecessary as importing manioc. Absolutely shameful.

The Science of the Flowers

When I've been in discussions about the legalization of marijuana, I have often run into the paradoxical argument that its
therapeutic use is banned because there has not been sufficient research to justify it, and also that this research cannot
be carried out because it's banned. The truth is that cannabis studies are nothing new, and long before prohibitionism
spread, a great deal of research was produced that did provide a basis for its many therapeutic uses.

In 1839, the Irish doctor William Brooke O'Shaughnessy published a detailed study of marijuana's medicinal
properties, carried out when he was serving with the British Army in India. In this study, O'Shaughnessy successfully
demonstrated the use of marijuana preparations for treating
convulsions, spasms, and rheumatism. His discoveries only
confirmed for the Europeans what traditional wisdom about
the plant on other continents had long known:

> The narcotic effects of Hemp are popularly
> known in the south of Africa, South America,
> Turkey, Egypt, Asia Minor, India, and the adja-

cent territories of the Malays, Burmese, and
Siamese. In all these countries Hemp is used in
various forms, by the dissipated and depraved,
as the ready agent of a pleasing intoxication. In
the popular medicine of these nations, we find
it extensively employed for a multitude of affec-
tions. But in western Europe its use either as a
stimulant or as a remedy, is equally unknown.[1]

O'Shaughnessy's study triggered a great deal of scientific
interest in marijuana, and even before the nineteenth century
was up, dozens of articles had been published on the subject.
European medicine began to make ample use of marijuana,
in the form of extracts, tinctures, and even cigarettes to combat asthma, catarrh, and insomnia. Huge pharmaceutical
companies like Bristol Myers Squibb in the U.S. and Merck
in Germany began to commercialize marijuana-based
products. A 1922 compilation listed marijuana's analgesic,
anxiolytic, sedative, and digestive effects as its main therapeutic uses.[2] A medical compendium in 1930 lists the following therapeutic properties of marijuana extract:

A hypnotic and sedative of varied effects, already known to Dioscorides and Pliny, its use
requires caution, the result being the good use
of this valuable preparation as a tranquillizer
and antispasmodic [. . .]. It is used for dyspep-
sia [. . .], for cancer and gastric ulcers [. . .], in
insomnia, neuralgia, mental disturbances [. . .],
chronic dysentery, asthma, etc.[3]

Despite the widespread recognition of its therapeutic properties, marijuana was progressively stigmatized in the 1920s and 1930s, until it was excluded from hegemonic medicine entirely. Unfortunately, Brazil played a part in this process, failing to oppose through the deceptive testimony offered by the Egyptian delegate Dr. Mohamed El Guindy at the Second International Opium Conference, convened by the League of Nations in 1924. Speaking in Geneva as a representative of a country with endemic marijuana use, El Guindy stated that marijuana was "even more dangerous than opium." This outlandish opinion was enthusiastically supported by the representatives from Greece, France, and Poland, without opposition from other countries like Brazil or India where marijuana use was prevalent. From that moment, international fora began to see a consolidating of the idea that marijuana should be universally banned.[4]

Unlike what many people suppose, however, this purge did not come about for any legitimate scientific or biomedical reasons but because of commercial interests opposed to hemp—the cotton industry and then, immediately after that, the nylon industry—and by racist political interests that were anti-Black and -brown in Brazil and anti-Black and -Mexican in the U.S. Entirely misleading propaganda movies and books began to be produced and disseminated using public funds.

In 1930, the U.S. created the Federal Bureau of Narcotics (FBN), and its director, Harry Anslinger, began to promote the idea of marijuana as a factory of perverts and criminals. In 1935, a poster produced by the FBN showed a joint with the words BEWARE! YOUNG AND OLD—PEOPLE

IN ALL WALKS OF LIFE! THIS MAY BE HANDED [TO]
YOU BY THE FRIENDLY STRANGER. IT CONTAINS THE
KILLER DRUG "MARIHUANA"—A POWERFUL NARCOTIC
IN WHICH LURKS MURDER! INSANITY! DEATH! Another
poster from the same period showed a woman wearing lin-
gerie and a nightgown and another who looked similar being
injected by a man with a syringe, alongside the words: MARI-
HUANA—WEED WITH ROOTS IN HELL. WEIRD ORGIES,
WILD PARTIES, UNLEASHED PASSIONS. SMOKE THAT GETS
IN YOUTH'S EYES. In the center of the poster, a devilish
hand is holding cigarettes with the labels LUST, CRIME, SOR-
ROW, HATE, SHAME, and DESPAIR. Beside them, a syringe is
labeled with the word MISERY.

The following year, in 1936, the movie *Reefer Madness* strik-
ingly enacted the official narrative about marijuana, presenting large audiences with the idea that consumption of the
plant leads to insanity and death. The movie shows young
people who are turned into marijuana "addicts" by seductive
dealers, causing them to spiral criminally down from negli-
gent bodily harm to rape, premeditated murder, and suicide.
With all this aggressively negative propaganda, it's no wonder the general public developed a great fear of marijuana.
The ideological ground had at last been prepared for the
1937 Marihuana Tax Act, which officially banned marijuana
across all U.S. territory.

Despite all its textile and therapeutic uses, which had
been known for millennia, during the 1930s the plant was
slandered, reviled, and ended up totally banned in Brazil, Mexico, South Africa, Canada, Indonesia, Thailand,
and the U.K., among other countries. In 1961, the United

Defamatory posters from the U.S. Federal Bureau of
Narcotics (Wikimedia Commons)

Nations signed an international convention to ban various substances, marijuana among them. For most people on the planet, marijuana came to be considered a terrible poison, a dangerous narcotic. This negative propaganda continued to be cultivated until a persistent moral panic had taken root.

But how did we begin to overcome this decline? The revolution that transformed the damned weed into an ever more socially valued medicine would not have happened without a corresponding revolution in scientific knowledge relating to marijuana and its incredible molecules. If ever the Nobel Prize recognizes the discoveries that nowadays form the basis for cannabis-based therapies, two brilliant scientists will doubtless be remembered as having deserved that recognition, even if they are no longer with us.

The Israeli Raphael Mechoulam (1930–2023) led the first global discoveries on the biological and psychological mechanisms of marijuana. In the 1960s, he identified THC as the plant's main psychoactive component.[5] Mechoulam was also involved in the pioneering discoveries of his friend and collaborator, the Brazilian Elisaldo Carlini (1930–2022), in his demonstration in the 1970s and 1980s of the anti-epileptic effects of CBD—first in mice, and then in humans.[6] These results were recently confirmed in rigorous clinical tests—controlled, randomized, and double-blind—in other words, by tests in which the patients were divided into random groups, and in which neither researchers nor patients knew which substance was being administered.[7] In the 1990s, Mechoulam and his team discovered anandamide, the first molecule identified as an endocannabinoid—that is, a cannabinoid that is endogenous, produced by the body itself, but capable of binding to the same CB1 receptor to which the plant-produced THC binds.[8]

After decades of suppression, biomedical research into marijuana and its components had finally taken flight. The year 1992, when Mechoulam and his team discovered anandamide, saw 123 studies on cannabinoids published across the world. In 2022, three decades later, 2,683 studies were published, a more than twentyfold increase.

Because of Mechoulam and Carlini's research, thousands more researchers all over the planet are today investigating the best way to use marijuana and its derivatives to treat a wide range of illnesses, as well as seeking to better understand the role of endocannabinoids in the maintaining of physiological equilibrium.

During my own doctoral and postdoctoral work, I witnessed this vertiginous rise in scientific interest in cannabinoids. At the first international neuroscience conferences I took part in, in the mid-1990s, almost nobody was taking any interest in the subject. The poster sessions on this topic attracted little notice, and the new field seemed no more than a curiosity, at best meriting a footnote in textbooks on metabolism or physiology. Before long, however, that interest would explode, and year after year, the subject occupied ever more space in those congresses' scientific programming. The endocannabinoid system had stopped being an appendix to textbooks and become an essential chapter on any biomedical teaching material that could aspire to be up-to-date. Nowadays, nobody doubts that an understanding of the endocannabinoid system is essential for the biological understanding not only of humans but of all vertebrates (the situation in invertebrates is not yet a matter of consensus).

And while some advances have already been made, others remain still underway. Various researchers have sought to understand the mechanisms by which cannabinoids reduce epileptic fits, spasms, and tremors. When cannabinoids are administered to lab rats, their neurons neither increase nor decrease their electrical activity, but they do begin to activate with a lower degree of synchrony, that is, they tend to fire just slightly before or after one another.[9] The practical effect of this reduction in synchrony in large neural populations is an inhibiting of aberrant or excessive brainwaves, such as those that occur during a convulsion or in the dyskinesia of Parkinson's disease, though without there being an inhibiting of the normal electrical oscillations that produce behaviors necessary for a healthy life. Similar mechanisms might be involved in the successful use of a CBD-rich oil to treat various symptoms of autism,[10] a condition that can also entail epileptic discharges.[11]

Yang, Yin, and Many Other Molecules Besides

In marijuana flowers, we can find more than five hundred molecules of biomedical interest, yet almost all of the scientific research carried out thus far has focused only on THC and CBD, the two molecules that are particularly abundant in most varieties of the plant. Though they have the same chemical formula ($C_{21}H_{30}O_2$), and a practically identical molecular structure, THC and CBD produce quite different effects on our metabolism, on our physiology, and on our psychology.

THC is a small, rigid molecule that powerfully activates the two main cannabinoid-receptor proteins in the body, CB1[1] and CB2.[2] These proteins are located on the surface of different types of cells, including neurons and cells in the immune system, and their activation causes a number of alterations to these cells' metabolism.

Compared to THC, CBD lacks a bond between carbon atoms. This small detail is enough to make CBD a more flexible molecule, less rigid, which cannot form a stable bond with the specific region of the receptor proteins to which THC binds. On the other hand, precisely because it is a more flexible molecule, CBD binds weakly to several different regions of various receptor proteins. Among other effects, CBD inhibits THC's capacity to bond with the CB1 and CB2 receptor proteins.

Marijuana's two main cannabinoid molecules, THC and
CBD, possess an almost identical molecular structure but
have practically the opposite psychological effects.
(Wikimedia Commons)

As a result, for most people, THC and CBD have opposite effects. While THC provokes euphoria and an acceleration of thoughts, CBD calms the mind and helps with the onset of sleep. Hence these two molecules that are typically most abundant in marijuana flowers explain Chinese medicine's traditional view, attributed to Emperor Shennong, that marijuana contains both yang and yin energies, that is, complementary principles that lend themselves to countless functions depending on the proportion of one to the other.

The physiological dichotomy between THC and CBD spawned the idea that the two main marijuana types, *Cannabis sativa* and *Cannabis indica*, would be particularly rich in THC and CBD, respectively. This dichotomy also led to marijuana being classified into three main chemical types: chemotype 1 with a predominance of THC, chemotype 2 with THC and CBD in balance, and chemotype 3 with CBD predominant. While this classification is very useful for research, for the pharmaceutical market, and for the adoption of public policies on scientific bases, actually marijuana's molecular reality is much more complex, interesting, and also promising, as it includes three distinct main families of molecules: cannabinoids, terpenes, and flavonoids.

Cannabigerol (CBG), for example, is a cannabinoid that has been attracting increased attention owing to its ability to reduce anxiety, chronic pain, depression, and insomnia,[3] with particularly intense effects on increased appetite[4]—which has a direct application in the treatment of the cachexia* that is frequently observed in patients undergoing chemotherapy.[5]

Another frontier of research into cannabinoids is the ever-more concrete possibility of using them to halt, or even reverse, the cognitive effects of Alzheimer's disease and other consequences of aging. Beta-amyloids are misfolded proteins that accumulate and cluster together within cells over time, and they seem to be aggravated by bad habits like poor sleep, eating ultra-processed food, and a lack of physical and intellectual exercise.[6] The toxicity of this protein clustering triggers an inflammatory response that can lead to cell death and, gradually over the years, the multiple cognitive symptoms of Alzheimer's disease.[7] In recent years, research at the Salk Institute in the U.S. has shown that THC and other cannabinoids such as cannabinol (CBN) can stimulate the removal of the beta-amyloid proteins from inside the neurons, blocking the inflammatory response and potentially protecting the brain from Alzheimer's.[8]

There are many other cannabinoid molecules awaiting

---

* Cachexia is the intense, pathological loss of adipose and muscle tissue, with a significant increase in inflammation. It is associated with a number of illnesses and is common with some types of cancer.

the attention of scientists, and recent evidence has suggested that the distinctions between the many varieties of *Cannabis sativa* and *Cannabis indica* depend even more on terpenes than on the cannabinoids.[9] Terpenes are the main constituents of the essential oils that provide such delicious scents, like the lemon and pineapple smell found in marijuana flowers and in many other plants, and that show various analgesic, anti-inflammatory, and antimicrobial properties.[10] One of the most common terpenes, the limonene found in citrus fruits, reduces the level of the inflammatory proteins called cytokines and presents scarring and antidepressant effects in animal models.[11]

We're still just at the start of an incredible journey of discovery about the therapeutic properties of the multiple possible combinations of so many different molecules. Imagine a sound desk covered in buttons. What is the total number of sonic possibilities? Countless . . . With that analogy in mind, we can understand why there are flowers that have such different effects. There are good flowers for laughing, for concentrating, for distracting, for sleeping or working, waiting or playing, for pleasure or childbirth. As with wine, the effects of the flowers depend not only on their genetics but also on the characteristics of their harvesting, maturation, extraction, and use.

It's important to know that marijuana in its natural state contains neither THC nor CBD but rather their acidic forms, THCA and CBDA, which are not psychoactive and which produce robust neuroprotective effects; that is, they delay or prevent the death of nerve cells.[12] When these acidic cannabinoids are heated above a certain temperature threshold, they undergo a chemical reaction of decarboxylation that converts them into THC and CBD. This is why marijuana is usually heated for consumption, whether during the extraction of the oil or by burning, vaporizing, or cooking the weed. Different temperatures during maturation, extraction, and use will affect the relative concentrations of THC, CBD, and countless other molecules, such as terpenes, which evaporate at relatively low temperatures.

All of this matters because recent research has corroborated the idea that marijuana molecules act cooperatively to produce different therapeutic effects, like the compensatory dynamic between THC and CBD. This molecular cooperation, dubbed "the entourage effect" by Raphael Mechoulam and his collaborators,[13] has been revealing an exciting new frontier in scientific research.[14] Beta-caryophyllene, for example, a terpene with a smell that is hot, woody, and spicy, amplifies the analgesic effects of CBD.[15] Other terpenes, such as alpha-humulene, geraniol, linalool, and beta-pinene, also show additive effects when combined with cannabinoids. A recently published clinical case study describes the successful use of a mixture containing various different terpenes—alpha-pinene, limonene, linalool, beta-caryophyllene, and nerolidol—to reverse the induced tolerance from three years of continuous treatment with CBD in an adolescent on the autism spectrum.[16] Induced tolerance is the loss of efficacy after some months of treatment.[17] And so, while marijuana oil that is rich in CBD and with a minimum content of THC is known to be effective for treating various kinds of epilepsy, at least in some cases it's necessary also to have the action of terpenes if this effectiveness is to be sustained.

In addition, epileptic patients and their families often report better results when they use oils that also contain a small
dose of THC. This is counterintuitive, since THC in isolation can have an opposite effect to that desired, worsening
the epileptic state. However, when used in combination with
CBD, and depending on the type of epilepsy, the role played
by THC might be different.

One of the leading specialists on the endocannabinoid
system in Latin America, the biologist Renato Malcher-Lopes
from the University of Brasilia (UNB), whose son Cauê was
diagnosed with autism spectrum disorder (ASD) back in early
childhood, sees in the entourage effect a concrete prospect
of a better life for the patient and his relatives. After thirty
months of treatment with only CBD, Cauê's father and his
mother Claudine Ferrão decided to add THC to the therapy
and were pleased at the effect. According to Renato, "I saw
a more powerful result. For my son's particular case, CBD
and THC in combination bring greater gains. It's important to emphasize that this is not a cure, but it is a stable improvement that acts on the reconfiguration of the brain."[18]
Cauê began to sleep more than seven hours a night, instead
of three or four hours, and his episodes of self-harm stopped.

The successful experience with Cauê motivated Renato,
Patrícia Montagner, and a team of researchers from a number
of institutions to carry out a retrospective analysis of twenty
patients with symptoms of ASD who were treated with cannabis oil containing a broad spectrum of cannabinoids, on
dosages that were individualized and based on the patients'
responses. Eighteen patients began with a protocol of ti-
trating oil rich in CBD, and for three of them, the CBD-rich

oil was gradually complemented with low doses of an oil that was rich in THC. The results were mainly positive for most symptoms: Eighteen out of the twenty patients presented improvements in the main ASD symptoms and comorbidities and in quality of life—their own and their families'—with only mild and infrequent side effects. In addition to this, the study showed for the first time that broad-spectrum cannabis oil is effective for treating the allotriophagy or pica that is typical of ASD, which consists of an irresistible desire to eat things that are not edible.[19]

The loving yet also scientific search for the most suitable treatment for Cauê was only possible thanks to Renato's intensive dialogue with doctors like Paulo Fleury and Leandro Ramires, who is also the father of a son with ASD and epilepsy and who at the time was president of the Brazilian Association of Medicinal Cannabis Patients (AMA+ME), one of the first such associations created in Brazil, in 2015.[20] The pioneering courage of these and other prescribing doctors helped to move public opinion toward something that science actually shows: Marijuana can successfully treat ASD, reducing or even eliminating attacks of rage and episodes of self-harm.[21]

If we think about cannabis therapy, which would be better? A single molecule drawn from the plant and purified, to be used in precise doses, or a phytotherapeutic extract with a broad spectrum of constituent molecules, for use in dosage ranges that are a bit flexible? The response depends on the particular case. There are undoubtedly clinical and pathological cases that require the former approach, such as certain epilepsies in which treatment with isolated CBD is the most suitable. However, there are many other situations, such as other types of epilepsy and many cases of anxiety, depression, or ASD, where the latter approach can be sufficient and even essential for avoiding tolerance and balancing out the euphoriant and sedative effects of THC and CBD, respectively.

Let us think, for example, of the use of analgesics for headaches. We don't step onto a scale each time to check our own body weight so as to know precisely what fraction of a pill we ought to take. What most people do is simply work within dosage ranges, corresponding to half, one, two, or more pills. Another useful example is that of vitamin C. We know we need to consume it regularly so as to avoid illnesses like scurvy, but we don't know the precise dose we consume each day via citrus fruits.

This question of dosage is important because the big pharma company lobbyists would have the public believe that it's only purified molecules in very precise doses that have a therapeutic effect and bear science's sacrosanct seal of approval. This is a malicious lie, as it is far more expensive and complicated to purify substances than to use their broad-spectrum extracts, which can be prepared at extremely low cost by patients' associations or even domestically.

The partnerships between growers, associations, universities, and research institutes can enable a truly diversified ecosystem for developing cannabis-based therapeutic products, with individuals, associations, and companies of all sizes occupying the various different niches in the market, promoting the broadest possible access to medications without ignoring the right to information about the doses of the different constituents, the expiry date, the degree of purity, any possible contaminants, and the risk groups.

Of course, the pharmaceutical industry has no interest in any version of marijuana-based therapy that does not bring them a royalty payment on its sale, which is why they try to make the phytotherapeutic approach impossible, accusing it of being ineffective or unscientific. It is neither. Fortunately, countries such as Canada, the Netherlands, Israel, and Uruguay have taken major steps in the clinical use of the flowers *in natura* and of the marijuana oil made from broad-spectrum extracts.[22] Besides the difference in cost, this is important because treatment with just the one molecule does not always sustain its effectiveness over time, as mentioned above.

The truth is that we are still moving at a crawling pace in our knowledge of the entourage effect and of the countless therapeutic possibilities created by the combination of different doses of CBD, THC, and the hundreds of other molecules present in marijuana, which interact with hundreds of molecules inside our bodies.[23] There are no simple dose-response relationships but rather complex interactions that connect drug (*substance*), body (*set*), and context of use (*setting*). The same is true for other substances, such as antidepressants and ayahuasca. What we do know for sure is that excessive focus on pure compounds will hinder a balanced perspective on the importance of the entourage effect and can make many patients' lives harder.

Marijuana Doesn't Kill Neurons, It Makes Them Flourish

The idea that marijuana is harmful to the brain and makes people lazy has spread across the planet. In Australia, in 2015, a negative ad campaign presented a user as a sloth—slow and inefficient.[1] In 2021, Brazilian police picked up a fake banknote that bore drawings of a sloth and a marijuana leaf, with a suggestive denomination: 420 reais[2]—an allusion to the universal time of marijuana consumption: 4:20 p.m.

It's embarrassing that the prohibitionists—typically out-of-date, elderly gentlemen including psychiatrists, police officers, and politicians—insist upon moral panic around marijuana, claiming that it causes indolence and low cognitive performance. In an attempt to stigmatize "potheads" as people who have chosen to be "soft in the head" and to live "in a state of laziness," they emulate the capitalist colonial criticism of the supposed laziness of Indigenous and African people who were violently enslaved. The police chief and Brazilian house representative Laerte Bessa was ordered to pay compensation of 30,000 reais to the former governor of the Federal District Rodrigo Rollemberg for having publicly called him "weak," "idle," "lazy," "incompetent," and—of course, there was no way he wasn't going to—"pothead."[3]

A challenge to these gentlemen and their prejudiced arguments: If we compare adult smokers of marijuana with a matched sample of whiskey drinkers or users of Prozac or Rivotril, which group will do better in short-term memory tests, access to long-term memories, cognitive flexibility, creativity, performance in sport, professional vitality, and sexual satisfaction?

Besides the prejudice against marijuana, the idea—so widespread now—that it makes you "dull" is rooted in a great deal of ignorance. Part of the problem comes from confusing short-term memories, which we use for temporarily storing a piece of information relevant for navigating day-to-day life ("Where did I leave those glasses?" "What time did we agree to meet?"), and long-term memories, which give us the store of integrated, interlinked experiences that we call the unconscious, from which our entire autobiographical narrative emerges—and our imagination, too.

In order to understand the difference, imagine your memories are stored in a backpack that grows over the course of your life, accommodating new experiences every day. During waking hours, new memories gather very close to the opening of the backpack, not yet well integrated with old memories, which are located deeper in the bag. Hours later, during our sleep, recent memories are merged in with the old ones, getting moved around and reorganized so as to optimize the use of space.

Working memories are temporary and they get discarded after use, so they never get very deep into the backpack.

Long-term memories, meanwhile, get deeper and deeper over time, and can remain in the backpack for the whole of a life.

It is well-known that, in isolation, THC causes a transitory deficit in working memory.[4] CBD, however, protects that memory[5] and can mitigate the harm caused by THC.[6] Inexperienced or occasional users of marijuana can experience cognitive difficulties, finding themselves unable to complete a sentence, which feeds the dumb-pothead stereotype. The consumption of flowers that are high in THC and low in CBD can lead to embarrassing situations, such as losing objects that are in full view or forgetting a subject discussed just moments previously. What most people don't know, however, is that those flowers that are rich in THC have a positive effect on long-term memory.

This surprising fact was confirmed very candidly by the multitalented Brazilian artist Nelson Motta, in an interview he gave in 2019 at the age of seventy-five:

> I've got an amazing memory, I don't know why. I've been smoking pot every day, for 55 years. Maybe because I started late, at twenty. They say when you start early, that's when it affects your neurons. My dad used to say I was living proof of that belief. Good to preserve that, isn't it? As people get older, the hard drive fills up.[7]

Over the course of our lives, as we sleep, our brain produces and selects new synapses (that is, connections between neurons).[8] However, with time, this capacity, called synaptogenesis, gets reduced. Something similar happens with the formation of new neurons in the hippocampus,[9] a region of the brain that is vital for the acquiring of new memories, as this capacity, called neurogenesis, reduces significantly after adolescence.[10] The explanation for the "elephant memory" that Nelson Motta described has to do with the capacity of THC and other cannabinoids to promote the formation of new neurons and new synapses.

An experiment carried out on adult mice confirmed a reduction in the capacity to produce new neurons in animals that were stripped of two cannabinoid-receptor proteins: CBD and vanilloid (VR1). In other words, these cannabinoid receptors play a key role in the regulation of neurogenesis.[11]

A subsequent experiment showed that the synthetic cannabinoid HU210 not only promotes neurogenesis in adult rats, but it also produces antianxiety and antidepressant effects that result from this increase in neurogenesis.[12] Curiously, this is the same mechanism that is supposed to explain the therapeutic effects of conventional antidepressants.[13] One way of understanding how the increase in neurogenesis reduces anxiety and depression is to consider that new neurons, by definition, don't remember the past. Their activation, therefore, will not contribute to the painful ruminating over memories that produces negative thoughts.

In a third experiment, also carried out on mice, Andreas Zimmer and other researchers at the University of Bonn in Germany demonstrated that adult animals given a moderate dose of THC for twenty-eight days experienced a significant improvement in cognitive performance.[14] The positive effects in tests of spatial navigation, object recognition, and the recognition of other individuals made it possible to bring the cognitive performance of old animals that were treated with THC level with that of young animals that were not. Curiously, in mice that were young, the opposite result appeared: That is, the treatment with THC worsened cognitive performance, similar to what had already been reported in studies on adolescent humans.

The researchers investigated the effects of treatment with THC on gene expression within the neurons of mice in order to understand the mechanisms causing these effects that are so variable according to age. To interpret the results of this research, it's important to remember that while all of the cells in an organism's body have the same genes, each cell uses— that is, it *expresses*—at each moment a very particular, very specific combination of its genome, its collection of genes. It's the differences in gene expression between the cells that grant them their morphological and functional differences, making some of them neurons while others are muscle cells, adipose cells, hepatic cells, renal cells, etc.

What Andreas Zimmer and his collaborators discovered is that the cognitive benefits of chronic treatment with THC in adult animals is mediated by the increase in the expression of those genes that promote the formation of new synapses, as well as the reduction in the expression of genes that promote aging. In the young, however, the opposite happens. It's as if THC rejuvenated the elderly and aged the young. Microscopic observation of the cell projections called dendritic spines, in which neuronal synapses are formed, confirmed that THC had the opposite effects in young and old animals. In older animals, THC improves the stability of the dendritic spines, while in young animals, it destabilizes them.[15]

The group of researchers then tested the hypothesis that these effects are a reflection of the natural decaying of the endocannabinoid system over time. This hypothesis was formulated more than twenty years ago, when the group discovered that mice that have been genetically deprived of CB1 receptors experienced a drop in bodily activity, an increased sensitivity to pain, and a high mortality rate, as if they were aging faster.[16] Subsequent studies showed a fall in the cerebral levels of CB1 receptors in old mice,[17] as well as a reduction in the levels of one of the main endocannabinoid molecules, 2-arachidonoyl glycerol (2-AG).[18]

The results obtained in rodents were compatible with those found in humans. The levels of the CB1 receptor are reduced in adulthood, from a peak reached between birth and early childhood. The levels of the enzyme that synthetizes 2-AG are shaped like an inverted U: They are extremely low at the beginning of life and in later adulthood, peaking in adolescence.

The idea that the aging of the endocannabinoid system leads to the progressive decay of the stability of the synapses and of cognitive functions led Zimmer's group to suggest that the endocannabinoid system suffers a "a middle-age crisis [. . .] which could be a potential time window for therapeutic interventions to abrogate the course of cognitive aging."[19] In other words, the administering of THC to elderly people can combat the adverse effects of the decline of the endocannabinoid system, roughly like what happens in widely used hormone replacement therapies, such as the administering of testosterone to women in menopause. CBD, on the other hand, reduces the efficacy of THC in the preserving of cognitive function in older animals.[20]

Taken together, these studies exemplify very clearly how false and simplistic the Manichean logic is, of God's substances and the Devil's, as if THC were a poison and CBD a balm, independent of their user's genetics, life history, and pathologies. CBD is not a substance "of good," nor is THC "of evil." Both molecules have considerable therapeutic and recreational application, with different and complementary uses. The studies also corroborate what Nelson Motta, that veteran connoisseur of fine flowers, expressed so astutely: Marijuana tends to do adult minds good.

It's worth reading Carl Sagan's detailed account of his first encounter with weed, published anonymously and its authorship revealed only after his death:

> I had become friendly with a group of people who occasionally smoked cannabis, irregularly, but with evident pleasure. Initially I was unwilling to partake, but the apparent euphoria that cannabis produced and the fact that there was no physiological addiction to the plant eventually persuaded me to try. My initial experiences were entirely disappointing; there was no effect at all, and I began to entertain a variety of hypotheses about cannabis being a placebo which worked by expectation and hyperventilation rather than by chemistry. After about five or six unsuccessful

attempts, however, it happened. I was lying on my back in a friend's living room idly examining the pattern of shadows on the ceiling cast by a potted plant (not cannabis!). I suddenly realized that I was examining an intricately detailed miniature Volkswagen, distinctly outlined by the shadows. I was very skeptical at this perception, and tried to find inconsistencies between Volkswagens and what I viewed on the ceiling. But it was all there, down to hubcaps, license plate, chrome, and even the small handle used for opening the trunk. When I closed my eyes, I was stunned to find that there was a movie going on the inside of my eyelids. Flash . . . a simple country scene with red farmhouse, a blue sky, white clouds, yellow path meandering over green hills to the horizon . . . Flash . . . same scene, orange house, brown sky, red clouds, yellow path, violet fields . . . Flash . . . Flash . . . Flash. The flashes came about once a heartbeat. Each flash brought the same simple scene into view, but each time with a different set of colors . . . exquisitely deep hues, and astonishingly harmonious in their juxtaposition. Since then I have smoked occasionally and enjoyed it thoroughly. It amplifies torpid sensibilities and produces what to me are even more interesting effects [. . .].

Looking at fires when high, by the way, especially through one of those prism kaleidoscopes which image their surroundings, is an

extraordinarily moving and beautiful experience. [. . .]

When I'm high I can penetrate into the past,
recall childhood memories, friends, relatives,
playthings, streets, smells, sounds, and tastes
from a vanished era. I can reconstruct the actual occurrences in childhood events only half
understood at the time.[21]

To summarize the message of this chapter, marijuana is
for adults and the elderly, not for children and adolescents.
Young people, unless afflicted by very specific pathologies,
ought to avoid marijuana. As we will see, premature use of
marijuana can lead to decreased motivation and poor school
performance. The more mature the user of these flowers, the
greater the benefits will be, and the lower the risks.

Living with the Flowers

Cultivating the Flowers

The verb "to cultivate" comes from the medieval Latin "*cultivare*," which derives from the perfect participle of the verb "*colo*," the possible meanings of which encompass the verbs "protect," "care for," "honor," and "adore." Through almost the entire Neolithic period, it was in the agroforestry, gardens, and orchards around the houses of the people—especially women—who did the cultivating where marijuana and countless other domestic plants were protected and cared for, creating the bases for agriculture, cooking, and pharmacology.[1] It was the systematic, intelligent selection of the sown seeds that produced the marvelous variety of plants we have at our disposal today, an inheritance of inestimable value that we must respect and use wisely. It is impossible to honor marijuana without honoring the countless generations of people who have cultivated it over these past twelve thousand years. Not only their physical labor but also their wisdom.

The exclusive focus on pure compounds produced by the pharmaceutical industry doesn't just lose sight of the entourage effect—as we saw in the fifth chapter, "Yang, Yin, and Many Other Molecules Besides"—but it also prevents a balanced perspective on the importance of the context of the treatment (*setting*), which can be critical for a successful cure. The traditional use of medicinal plants tends to place great importance on the context of the treatment for maintaining health and arriving at a cure, through practices that strongly mobilize the desire to attain it. This process proves to be a positive placebo effect—in addition to the pharmacological effects—which is caused by the context of the therapeutic practices and the way in which the patient interprets them.[2]

This idea seems counterintuitive, since common sense has come to understand the placebo effect as something undesirable because it is "fake," as if it could not produce genuine effects. On the contrary, we know today that the placebo effect can effectively reduce brain activity in areas associated with pain and negative emotions. The positive effects of a placebo on pain and emotion can be very useful in cases of depression or Parkinson's disease, for example, through the activating of motivational systems in subcortical regions.[3]

Gardeners tend to be calm, thorough, and constant. In the case of marijuana, the fact that it was banned added an extra necessary quality: courage. When the mothers and fathers of patients like Clárian, Sofia, and Anny decided to face up to the legislative obstacles to treat their children with marijuana oil, it was often the secret growers who rescued them, supplying for free the necessary remedy, without which life would simply wither away.

It's such a pleasure to see the respect, care, and even adoration these people show their plants. Holders of green-fingered mysteries, they know their cycles and are able to extract from their development the best flowers of good, filled with elated resinous trichomes, the epidermal glands from the flowerings of female plants, which secrete cannabinoids in high concentrations. They know that the plant's other parts, from the leaves to the roots and the seeds, all have incredible therapeutic and nutritional properties to be explored.[4] And they also know that hemp, with its deep roots, can be used successfully for extracting heavy metals and other contaminants from the ground and from the water.[5]

To all those who did and do this work for humanity, from the end of the Ice Age to today, a heartfelt thanks is due. At the same time, it's good to know that all these people did and do receive the direct benefits of their practice, as gardening is itself inherently good for the health, improving mental well-being, increasing physical activity, and reducing social isolation.[6] In the words of psychiatrist Oliver Sacks: "In forty years of medical practice, I have found only two types of non-pharmaceutical 'therapy' to be vitally important for patients with chronic neurological diseases: music and gardens."[7]

There is still much to be clarified about the potential in the loving relationship between marijuana and its growers, which in countless respects resembles the relation of mutual fidelity between dogs and those who care for them. What is

the effect of carrying out cannabis treatments with flowers
that were planted, cared for, monitored, harvested, mani-
cured,* cured, and processed by somebody who adores them?
Asking what the therapeutic synergy is between flowers and
their growers is equivalent to determining what the positive
placebo effect is of establishing a loving relationship with the
plant from which the curing medicine is extracted. Science
of the future will tell us.

Eating with the Flowers

One of the most characteristic effects of the consumption
of marijuana flowers is an increase in the appetite, involving both the enhancing of hunger and the refining of the
palate.[8] The familiar feeling of "the munchies," a sudden
hunger experienced by marijuana users, reflects the action
of THC, CBD, and possibly other endocannabinoids on the
hypothalamus, which through the endocannabinoid system
promotes energy balance and stimulates the appetite.[9] In his
personal testimony about marijuana, Carl Sagan says that
with it, "The enjoyment of food is amplified; tastes and aro-
mas emerge that for some reason we ordinarily seem to be
too busy to notice. I am able to give my full attention to the
sensation."[10]

The power of marijuana to affect the appetite is not limited to humans. In rats, THC increases the hedonic response

---

* Manicuring consists of removing the leaves that are close to the
  inflorescences.

to saccharose without increasing the aversive response to unpleasant tastes. The effect is mediated by the increase in the release of dopamine in the nucleus accumbens, a region of the brain involved in reinforcing pleasurable behaviors.[11] Experiments on mice showed that THC also prevents the process of becoming accustomed to pleasant smells, that is, animals treated with THC do not lose interest in these smells even after prolonged exposure. This happens because the THC activates receptors in the brain's olfactory bulb, sharpening the sense of smell and improving the taste of foods. As a result, the rodents keep on smelling and eating for much longer.[12]

The intimate relationship between marijuana and appetite makes the plant a powerful ally in overcoming clinical cases of malnutrition and cachexia. In promoting the healthy balance between energy expenditure and food ingestion, cannabinoids act directly on bodily regeneration. Beyond any strict therapeutic usefulness, marijuana sharpens the sensorial experience of taste. It is no accident that the U.S. and Canada are seeing a flourishing of cannabis gastronomy, which combines hunger with a sheer desire to eat.[13] As legislation advances, more and more people are discovering an ancestral secret: Curing with flowers can be as tasty as eating them, or eating other foods after having them.

Sweating with the Flowers

Anybody who thinks marijuana and physical activity don't mix doesn't know much—about one or the other, or about either. The use of cannabis in the sporting arena is an open secret among athletes in sports as varied as swimming, surfing, skateboarding, and basketball. Research on the reasons mentioned anonymously by people who exercise under the influence of marijuana encompasses the interviewees' physiological, psychological, neuromotor, and even spiritual spheres.[14] The swimmer Michael Phelps, the greatest Olympic medalist of all time, is a well-known user of marijuana. Why should that be? The truth is that, before or after training or competing, different varieties of flowers, with their anti-inflammatory and antioxidant properties, can favor muscle regeneration and relaxation, reducing pain and anxiety.

The orthopedist and pain specialist Dr. Ricardo Ferreira, one of the pioneers in prescribing cannabis in Brazil, states:

> Someone who is a high-performing athlete engaged in high-performance physical activities will already recognize the value of cannabis, both where it comes to injury prevention, and in terms of the recovery from physical exercise, and pain management. [. . .] Many athletes with a history of long-term high consumption, from the generations that came twenty or thirty years ago, already reported the use of cannabis as a way of improving their physical condition. [. . .] Unquestionably it greatly improves an athlete's quality of life and potential longevity. Even if science has not managed to completely explain how this happens, this has existed empirically for generations of athletes who use cannabis as

medication, especially in countries where access is easier. [. . .] An athlete will be predisposed to feel pain related to inflammation, and they know they're going to do something that will cause this. For example, a runner knows there will be some impact, that they'll be putting an excessive burden on their knee, their ankle, their calf, or on the thigh itself, so in other words, in anticipating this, their use of cannabinoids can be of interest as a way of preventing the pain. So you do this preventative thing—before you have the pain, you prevent it. Whereas if we talk about preventing [pain] through the medication that we have in pharmacology, in the pharmacy, such as anti-inflammatories, such as opioids, that would be unthinkable, because the negative effects of these substances in the medium and long term exceed the benefits.[15]

In 2022, the World Anti-Doping Agency (WADA) announced that they would review marijuana's position on their list of banned substances. In pro basketball in the United States, its use is so widespread that the powerful National Basketball Association (NBA) came to an agreement with the players' association simply to stop testing for marijuana use in their anti-doping regimes. Even more importantly, they abolished any punishment in the event of detection. A similar trend can be seen in the professional leagues of American football, hockey, and baseball.[16] The moral panic over marijuana is cooling off.

These winds of change have been allowing ex-NBA players like Matt Barnes to open up about their own particular way of benefiting from the flowers: "I would smoke cannabis six hours before a game. We'd have a shoot-around in the morning, I'd come home and smoke a joint, take a nap, shower, eat and go and play."[17]

Depending on the variety of flower chosen, on the dosage and on the gap between the use and the physical activity, it is possible to give oneself over to restorative sleep or increase the motivation to sweat. It is important, however, to stay away from high doses *during* the physical exercise, as marijuana can increase heart rate and blood pressure, heightening the risk of arrhythmias and strokes.[18]

Exaggerations aside, the global trend is toward the end of sporting sanctions against marijuana, as well as the end of hypocrisy. I'll never forget a serious conversation I had more than fifteen years ago with a marathon runner I would often spot on Cotovelo Beach in Rio Grande do Norte, when I was running along the seaside:

"Say, tell me something, my friend . . ."

"Sure."

"So you people, you marathon runners, who get all that inflammation in your joints running as much as you do . . ."

"Twenty-six miles."

"Right—well, then—how many of you use marijuana to help with motivation, pain, or recovery?"

"Nobody."

"Are you serious?"

"Yep—nobody."

"Come on, buddy. Tell me."

"No one. I'm serious."

"Really?"

"No . . . everybody."

"Hahahahaha!"

Loving with the Flowers

If there's a kind of cardiovascular exercise that could be one's last—which wouldn't be so bad, after all—it's sex. Saying that marijuana is an aphrodisiac doesn't do justice to the enormous range of possibilities that it can create for increasing the connection, communication, and enjoyment of sex, until the attainment of the deep and nourishing orgasm that (almost) everybody needs for a better life.

Science corroborates marijuana's erotic potential. A Canadian study on the effects of consuming weed before sex showed that most of the 216 participants reported an increase in relaxation, in sensitivity to touch, and in the intensity of their feelings, resulting in a more pleasurable sexual experience. Some people, however, reported sleepiness and a lack of focus, while others experienced no change at all. Still, 65 percent reported an increased intensity of their orgasms.[19]

A recent study of 2,790 participants from the Brazilian university community during the COVID-19 pandemic showed that the use of marijuana during that period improved their satisfaction with their sex lives.[20] Another study, likewise recent, carried out with 811 participants in the U.S., showed more than 70 percent reporting increases in desire and in the intensity of their orgasms, as well as heightened pleasure during masturbation and an improvement in the senses of touch and taste.[21] The increased libido caused by marijuana is greater with small and medium doses than in large ones.

There is still much to be (re-)discovered about the different ways of using marijuana to love better. The use of marijuana-based vaginal lubricants—popularly known as "xapa xana" ("stoned pussy")—has caused real sexual and emotional revolutions in women's lives. In India, marijuana has been used for sexual purposes since the Bronze Age, for its ability to induce a nonstop trance flow in the mind. The use of bhang, marijuana leaf tea, to facilitate this special type of sexual trance, is thousands of years old. In tantric sex, people seek the illumination of their consciousness through transcendental meditation during the sexual act, with a view to achieving the fullness of their ecstasy with a delaying of ejaculation and of the passing of time itself.

The intimate relationship between the pleasurable effects of marijuana and changes to one's perception of time did not go unnoticed by Carl Sagan:

> Cannabis also enhances the enjoyment of sex—on the one hand it gives an exquisite sensitivity, but on the other hand it postpones orgasm: in part by distracting me with the profusion of images passing before my eyes. The actual duration of orgasm seems to lengthen greatly, but this

may be the usual experience of time expansion which comes with cannabis smoking.[22]

In order to love fully, one must forget entirely about time, work, and death, to surrender oneself entirely to the present moment.

Time with the Flowers

Perhaps the most mysterious of all the powers to be found in the flowers of marijuana are the very peculiar alterations they cause to one's perception of the passage of time. A fifth-century Taoist priest wrote that cannabis was used by "necromancers, in combination with ginseng* to move time forward in order to reveal future events."[23] Under its effects, depending on dose and variety, time can turn liquid and viscous until it has almost frozen, or converge toward infinity without slowing. At excessive doses, time can become disordered, delocalized, out of gear; it can fade away or stop entirely. If all is well, then there will be time enough for everything, but if things are dangerously out of place, if there's something strange in the air, time can slip by and drain away in a moment, or even swirl around the world, spinning. Relaxed or suspicious, happy or concerned, future or past, time in the company of marijuana is a unique state of being. Profoundly rooted in the present, it becomes obvious: The present is all there is.

The scientific research that has been carried out on the

---

* Interestingly, ginseng is rich in terpenes.

subject shows that marijuana tends to provoke an overestimation of time.[24] A study of fifty regular users of cannabis and forty-nine nonusers suggested that the users' internal rhythms were reduced after marijuana was taken.[25] Another study on the acute effects of THC, designed as an experiment that was double-blind, randomized, counterbalanced, and controlled with a placebo, showed in forty-four individuals that any dosage prompted an overestimation of time—that is, an impression that more time had passed than really had. However, while occasional users or nonusers present with a temporal overestimation in medium and high doses, frequent users did not show significant changes with any dose.[26] This result calls to mind a line attributed to the great master of Capoeira Angola, Mestre Pastinha: "What I do smiling, you don't even do angry." Each person is different, and what for a beginner is very difficult or even impossible, to the initiated can be easy and hugely pleasurable.

Many people report an increase in attentive focus under the effects of marijuana, while others report the opposite, an increase in distractedness. Science doesn't yet know much about the complex relationship between perception of time, hyper-focus on what is of interest, and a lack of attention for what is not. A recent meta-analysis shows that people diagnosed with attention-deficit/hyperactivity disorder (ADHD) can show either improvement or deterioration of symptoms such as impulsiveness, trouble concentrating, or forgetfulness under the effects of marijuana,[27] but the research still needs to be developed further because the concentration and quantity of THC and CBD used were not carefully measured in most of the studies, nor was the patients' genetics characterized.

What some find confusing and obstructive, others experience as clarifying and constructive. Each person is him- or herself.

The journalist and writer Otavio Frias Filho gives a good description of the creative and emotional equilibrium that some find in marijuana:

> Everybody has centripetal and centrifugal aspects to their personality. I have that centripetal element very strongly. And marijuana gives me, or releases in me, the centrifugal element, which gives me a bit more liberty to think more freely, to speak more freely. So it's as if marijuana were positioning me in the middle ground where all individuals ought to be, ideally. Because it really does release this centrifugal part of me. And since I am otherwise so centripetal, it doesn't derail me, it doesn't make me too disorganized. I can still read and write, and remember what I've read.[28]

We see a similar testimonial from Nelson Motta:

> It was something that went beyond well-being and freeing the mind, for me it was something that brought people together, it was also something forbidden, there was something transgressive that brought me pleasure, and another thing started then that pleased me even more, which was productivity. It was really something starting to write while high! And you had the option of

revising it the following day. It just got better and better. [. . .] Working is my hobby so I like getting up early, having a good breakfast, lighting up my joint and getting started. Then my head's really fresh, everything's working well, I have my ideas, some good, some not so good, no matter, and that's how I started to do things.[29]

For somebody who is experienced in using marijuana to carry out a certain task, sustained practice ends up constructing an unconscious flow of those movements that are strictly necessary—and them alone. Like the Zen archer, it's simply an action well done, with no time to miss, no time to look back.

Free Canine-nabis!

Tragically, in Brazil, as in the United States and other countries, there are many poor Black people incarcerated for using or selling marijuana, while whiter, richer people tend to be spared this problem. A plant gift from our ancestors has become the key to the state's war against its most vulnerable. As the writer Sebastian Marincolo says, "The legalization of marijuana is not a dangerous experiment—the prohibition is the experiment, and it has failed dramatically, with millions of victims all around the world."[1] This is why in 2015, the Brazilian Supreme Court began to consider Extraordinary Appeal no. 635659, concerning the decriminalization of the possession of drugs for personal use. In June 2024, the court decriminalized marijuana possession of up to forty grams of flowers or six female plants.

One of the most brilliant musicians of all time, the revolutionary trumpeter and singer Louis Armstrong (1901–1971), spent his whole adult life bravely facing up to the racist repression of the weed he loved to smoke and to share with his friends. After overcoming all the obstacles of a childhood in extreme poverty and almost total abandonment by his father,

Armstrong left school at age eleven and started playing the trumpet and singing on the streets of New Orleans in exchange for a few coins. At seventeen, he was hired to join a band traveling up and down the Mississippi on a steamboat. On this floating music school, young Louis learned to sight-read and to play solos on his horn. A few years later, his abundant talent took him to New York and then to Chicago, where he had his first contact with marijuana.

It was love at first sight. He said: "It's a thousand times better than whiskey . . . It's an assistant—a friend."[2] The relaxing weed imported from Mexico was conducive to a good mood and to improvisation, the trademarks of a new musical genre that was beginning to spread very energetically across the United States: jazz. With marijuana, this creative, entertaining trumpeter became more creative and entertaining still, composing and recording a series of songs that were extremely well received by audiences—including the track "Muggles," a code name for marijuana, apparently the first recording of a musical improvisation, in 1928.[3]

Armstrong so liked weed that he insisted it be taken by the musicians in his bands before their recording sessions, so that they all be in playful harmony during the performance of melodies that were ever more liberated from written scores. Artists like Billie Holiday, Dexter Gordon, Cab Calloway, and Bing Crosby sought inspiration and joy in the lively smoke-filled parties hosted by "Pops"—a nickname he earned in reference to being the "father of marijuana," for his leadership in blowing the minds of so many artists.[4]

Crossing racial and class barriers, Armstrong was one of

the people most responsible for bringing the music produced in bars and nightclubs almost exclusively frequented by a Black and Latino clientele into the theaters, record companies, and radios that could reach white audiences on both sides of the Atlantic. Over the course of the 1920s and 1930s, the bold, smiling kid from New Orleans would become the king of jazz, the first Black person to achieve global acclaim, long before Pelé or Muhammad Ali.

At a time when racial segregation was law, and the lynchings of Black people were regular occurrences, jazz and marijuana played an essential role in resistance to horrors and to a reduction in social tensions. In Pops's words, marijuana "makes you feel good, man [. . .]. It relaxes you, makes you forget all the bad things that happen to a Negro. [. . .] [It's] an insulator against the pain of racism." But Armstrong's success did not insulate him from human racism, nor botanical racism. Pops was arrested for marijuana use in 1930, and in 1948 the FBI opened a secret file on him. Yet the legendary jazz musician never stopped defending the legalization of the plant. After all, in the words of the actor Bill Murray, "The most dangerous thing about weed is getting caught with it."[5] The police harassments experienced by Armstrong look a lot like those experienced many decades later by other Black musical geniuses, such as Fela Kuti, Gilberto Gil, Mano Brown, and the band Planet Hemp. Despite their fame and their success, all these artists suffered the racist persecution of the war on marijuana.

It's such a grim irony, the creating of so much fear and stress around the consumption of a plant that's so calming

and relaxing! Marcelo D2, Planet Hemp's founder and vocalist, recalls his feelings when he smoked his first joints, more
than three decades ago:

> I'm always trying to get back to them! To the
> highs of my first beques . . .* I remember them as
> moments of introspection, of thinking about life.
> I see myself as a sensitive kind of guy, who can
> feel what's around him, and I think marijuana
> enabled this in me. I never studied much, didn't
> have much access to culture, I only started feed-
> ing myself culture and knowledge when I started
> earning money and being able to travel. In that
> small world where I lived, marijuana gave me
> something like it. Marijuana was even like ther-
> apy! I didn't have the money to pay a therapist, so
> it was just smoking and thinking deep thoughts,
> great chats with my friends.[6]

Why prevent the working population's access to this ancestral medicine? In the words of Mano Brown, vocalist of
Racionais MC's:

> Those kids around these lanes and alleys, living
> those crazy lives, riding motorbikes all day long,
> y'know, that stress, your arms stiff, shoulders
> stiff, all day, risking their lives every moment,
> cause that's what being a biker is: risking your

---

* Another word for joints, marijuana cigarettes.

life on every bend, every moment [. . .] then you
get to that point in the day where all you want to
do is light up that joint, sit under a tree and say
"Oof—I'm alive."[7]

To global rejoicing, the war on marijuana started to end
in the very place it was invented, in the United States. Since
2022, the federal health authorities have been recommend-
ing that the DEA loosen its restrictions on cannabis and
reclassify it from Schedule 1, a classification that includes
those drugs considered highly addictive and with no therapeutic function, to Schedule 3, joining lower-risk substances
that are allowed to be researched without restrictions.[8] The
year 2022 was also the year when President Joe Biden par-
doned all those people incarcerated at the federal level for
dealing in or possessing marijuana. Before these people
sue the state for the various harms they have suffered, the
government has been trying to repair the damage, giving
former inmates preference in the granting of licenses to
market marijuana.

In different parts of the country, a political negotiation is
proceeding that could see some part of the taxes derived from
the commercialization of marijuana directed to programs of
reparation for those who suffered from its persecution. This
is in convergence with the broader agenda of anti-racist reparation. Remarking on the courageous legislative propos-
als for racial reparation in the city of San Francisco, Justin
Hansford, professor at the Howard University School of
Law, stressed the need for redistribution of resources in order
to correct things: "If you're going to try to say you're sorry,

you have to speak in the language that people understand, and money is that language."[9]

In Brazil, we're a long way from that yet, but activists and organizations like Rede Reforma are proposing similar mechanisms, such as directing the receipts from taxes on cannabis toward the communities in the periphery who suffer the worst impact of the war on drugs, and incentives for cannabis start-ups in those same places. The favelas in the large metropoles aspire to be a part of a diverse ecosystem of micro- and small firms, each specialized in one particular type of flower, a tropical terroir without terror or armed vehicles of police raids. Can you imagine rooftops covering the view with green, when you land in Rio de Janeiro and drive down along the side of the Favela da Maré? Can you imagine the bank of the Tietê greened over by enormous marijuana plantations?

In order to free canine-nabis—my nickname for the cannabis that our ancestors domesticated—we need to look at the plant as a wonderful gift from those who came before us, without stigma and without prejudice. Science has been gradually knocking down the countless myths that were created around the plant in order to demonize it. We have already seen how the popular notions that "marijuana kills neurons" and "marijuana is comparable to jararaca venom" are false. Another such lie is that marijuana is a gateway for the consumption of other drugs.

In the most recent and complete piece of research into drug use in Brazil, conducted by Fiocruz in 2017, 66 percent of the Brazilian population aged between twelve and sixty-five said they had consumed alcoholic drinks at least once in their lives, with around 30 percent having consumed at least one measure of an alcoholic drink in the thirty days preceding the research.[10] As for having smoked at least one mass-produced cigarette in their life, this was reported by 33.5 percent of the population aged twelve to sixty-five. In comparison, only 7.7 percent of the population in the same age bracket said they had consumed marijuana at least once in their life. If marijuana were the gateway drug to the others, its use should be more prevalent than that of alcohol and tobacco, which is not the case.

Studies carried out in the United States comparing the potential for dependency* make it clear that marijuana is persecuted unfairly. While users of nicotine show a probability of developing a dependency of 67.5 percent after using the drug on just one occasion, this value is 22.7 percent for users of alcohol, 20.9 percent for users of cocaine, and 8.9 percent for users of marijuana.[11]

Marijuana is definitively not the gateway into other drugs, but it can in certain cases be the exit route. In an observational clinical study carried out with problematic crack users, the psychiatrist Dartiu Xavier da Silveira and his collaborators at Unifesp confirmed that 68 percent of their participants reported using marijuana to reduce the acute symptoms of crack withdrawal.[12] The study pointed to the need to carry out a rigorous experimental intervention, that

---

* It's important to emphasize and problematize the concept of dependency here. Only a proportion of those users considered to be dependent—those who develop a habit—present symptoms of problematic use. More broadly, the problem is not in the person or in the drug but in the particular relationship between the two.

is, a randomized placebo-controlled clinical trial, to explore the possibility of using marijuana as an exit route for crack addiction. Unfortunately, prohibitionist colleagues at the same institution threatened to call the police if the follow-up study was begun. This bizarre situation was reported in the press at the time, but twenty-four years have since passed and to this day the study has not been carried out in that department.[13] However, in 2024, another research group led by Dr. Andrea Gallassi from the University of Brasília showed in a double-blind, randomized clinical trial that CBD was better for treating a crack addiction than fluoxetine, valproic acid, and clonazepam.[14]

Another myth that needs to be debunked is that marijuana causes schizophrenia. It is true that the odds of cannabis misuse are higher for people with psychosis than for the general population.[15] It is also true that marijuana can prompt psychotic breaks in people with a family history of psychosis. However, inferring a simple causal relationship between marijuana use and psychosis is difficult, as its consumption could in reality be reflecting a self-medication to mitigate the symptoms of psychosis or the adverse effects of their actual antipsychotic medication.[16] If marijuana were indeed a factory of schizophrenics, as some prohibitionist psychiatrists continue to maintain, the proportion of people with psychosis in the general population—under 1 percent[17]—should have grown considerably in past decades, along with the growing global use of marijuana[18]—and this has not happened.

The difficulty in establishing causes and consequences

between cannabis use and psychosis is due to the fact that marijuana can have opposing effects depending on its chemical composition. CBD has an antipsychotic effect that has been well established over three decades of research at the University of São Paulo at Ribeirão Preto,[19] which was subsequently replicated in other countries, such as Poland, Romania, and the United Kingdom.[20] Whereas THC, taken in large doses and in the absence of CBD, can indeed induce psychotic symptoms in certain people.[21] This is why it is important to map out the origin of people's vulnerability to THC. People with a genetic predisposition toward schizophrenia, such as those with variants of the gene for the catechol-O-methyltransferase enzyme,[22] which metabolizes neurotransmitters, or certain variants of the CB1 receptor,[23] can be quite susceptible to THC and need protecting via reliable information about this risk and about the presence of THC in marijuana.

Schizophrenia, however, goes far beyond the occurrence of a transitory psychotic episode, as the disease in which the episodes take place with increasing severity depends mostly on the patient's family genetics.[24] Most of the adolescents who abuse marijuana do not develop schizophrenia, just as most people with schizophrenia never used marijuana as adolescents. Could it be that individuals who are genetically predisposed to schizophrenia are self-medicating with marijuana? The largest genome-wide association study carried out to date on the lifelong use of marijuana, with 184,765 participants, suggests that this is the case.[25] Research has shown a positive causal influence of schizophrenia on cannabis use.

In other words, it is not generally cannabis users who develop schizophrenia but people with an increased risk of schizophrenia who seek out cannabis. As science progresses, the attempts to demonize marijuana are becoming ever weaker.

Maturana, Marijuana, and the Green Frog

I began this book by telling the story of how marijuana was demonized in my own family, when my brother and I were still teenagers. It is time to pick that story back up again. During the most acute period of family conflict, my relationship with marijuana did try to change, but it hit the post, and the change never actually happened. Sitting around a campfire on the shore of Lake Paranoá, a friend of mine and my brother's produced a joint. Júlio wasn't there. Curious, and with no sense of having anyone watching me, I inhaled for the first time. I didn't feel any perceptible alteration to my state of consciousness, just a subtle sense of relaxation, my mouth dry and eyes red.

This frustrating experience would be repeated a few times in that final year of high school and the start of my time at university, with no noticeable effects of any interest. I began to lose curiosity and finally persuaded myself that marijuana was all but ineffectual to me. In spite of this, I never had the nerve to tell my mother about these experiences. My elder-son persona continued to align with the family mantra that all the problems with my brother had been caused by his marijuana use. I moved through the degree course in biology at UNB, deep into my research and the student movement, disconnected both from my brother and from the weed. At that time, I wouldn't even notice if a joint showed up at a party or a gig—and when I did get invited to share it, I would decline.

My alienation from the plant changed radically in 1992. Finding myself deep in a vocational crisis concerning microbiology, the area I'd been researching since 1988, I decided to suspend my final undergrad semester to go backpacking alone, with barely any money in my pocket and no properly planned itinerary. I had set out with an ambitious but vague plan, to cross Argentina to Chile in order then to board a merchant ship that was leaving Valparaíso bound for India, but by the time I'd finally made it across the Pampas and the Andean mountain range, after two months of couch surfing or even sleeping under the stars, suffering endless hours at the roadside in the incessant wind, I had no energy to cross the Pacific Ocean or to explore Southeast Asia. There were no cell phones, there was no internet, and I was far from home so thought it would be better not to push it. I decided instead to travel through South America and try to reach Mexico (which never happened either, actually, but that's another story).

The first stage of this new route was traversing the whole length of Chile, from Patagonia in the far south to the Atacama Desert in the far north. Committed to this plan, I stuck out my thumb at the southbound highway exit at Santiago de Chile, and after two days and a few mishaps along the way, I was boarding the ferry that connects Puerto Montt to the city of Castro, on the island of Chiloé.

It was on that rainy Chilean island, covered in little yellow flowers, that I had my first significant encounter with marijuana. I was staying with friends of friends, very supportive and progressive people who did volunteer work to help the Indigenous Huilliche—or Williche, the "southern people" of the Mapuches who lived on the island—each of them using their own particular expertise: Ana María Olivera was a lawyer, Manuel Muños Millalonco an archaeologist, and Martín Correa a historian. They were all a little older than me and their actions were very much guided by the thinking of Humberto Maturana and Francisco Varela, two Chilean neurobiologists who posited the idea of circular causality between the brain and the world, that is, the idea that internal changes to the brain are as much cause as consequence of external environmental changes, via an unending cognitive flow.[1]

The possibility of understanding the brain as a dynamic system that is bidirectionally coupled to external reality was an idea I found seductive, just as I did the honest and loving attitudes of my hosts, who in their fraternal welcome never even asked me how long I was planning to stay. Manuel and Ana gave me the chance to do a bit of voluntary work on the excavation of a traditional Huilliche burial site. Before I knew it, I was on my knees on the damp ground, carefully brushing the last bits of clay off the side of a canoe that served as a coffin for an Indigenous elder. As I worked, I pondered the circularity of the brain-world relationship proposed by Maturana and Varela . . .

That day, we returned home as a full moon was rising from the horizon. I was euphoric at the huge number of new ideas I had in my head, immersed in the adventure of life, grateful for the stimulating company of these new friends who had so much to teach me. In the little stilt house where Martín lived, we gathered around the fire and shared wine, bread, and olive oil. We chatted, laughed, and sang. At one point, Ana lit up a joint to pass around the group. This time I didn't say no. The cigarette moved from hand to hand, and finally it was at my lips. I inhaled the smoke, held it for a second, and shut my eyes.

BHANG! For the first time, I was into a whole new mental state, which experienced users call a "trip." My thoughts sped up vertiginously and I felt that I no longer could—and no longer wanted to—control their flow. I was a little scared but soon surrendered myself to the whirlpool and at last was swallowed up entirely.

The first few moments were difficult, a feeling of genuine mental confusion, like I was tumbling into the void. I opened my eyes, scared, and found everything just where I had left it. The people were continuing to chat and laugh and dance. I calmed down. Though I couldn't quite understand what they were saying, I felt that all was well. My body was very relaxed now, I felt protected, and once again I closed my eyes.

BHANG again! Now I was no longer tumbling from a great height but coming down a slide of interesting ideas that passed quickly through my mind without my being able to grab ahold of any of them. I got used to the feeling of producing—and then allowing to slip through my fingers—a plentiful rosary of beautiful truths, a torrent of new discoveries in a process of constant dissipation. I resisted the desire to open my eyes, and relaxed still further, and noticed that time was slowing almost to a stop. I held back my fear again, I remembered to breathe and tried to see something in the darkness of my mind . . .

I kept my eyes closed for a period of time that was outside time and finally I began to see something in the dark. Very faint at first, but gradually the outlines of a huge internal space started to become clearer. As in a waking dream, I began to see my own thoughts as electrical activity spreading across neural networks, as if they were glowing cars moving through the city at nighttime. Each mental state corresponded to changes in the phasing of the traffic lights. For the first time, I had a vision that was at least slightly convincing of the biological mechanisms underlying my own mind, an emotional experience, in the first person, of the electrical, cellular, and molecular gears that produce ideas, dreams, and desires. Finally I fell asleep, comforted by the very clear realization of having experienced profound revelations.

When I woke the following day, I felt bonded to my Chilean companions, in a circle of protection and trust. On the other hand, I did have Cartesian doubts about all the previous night's neuropsychological revelations. If that illicit plant really could generate such interesting ideas, a lot of people would know about it and it would be widely used by scientists, inventors, artists, and creators in general, which was not the case (or so I thought).

Skeptical of the possibility that marijuana might be a factory of useful thoughts, I formulated the theory that the altered consciousness it provoked merely gave a veneer of truth to any idea of those who consumed it, independent of whether the idea is actually a good or bad one.

So I decided to carry out my first self-experiment with marijuana. I wrote an obvious fact on a bit of paper—"the frog is green"—and decided I would read these words the next time I had contact with the plant. No sooner said than done: The following night the gathering around the fire happened once again, and the weed was again shared, and when I read the dumb words written on the bit of paper, I once again had the intense feeling of a great truth revealed. Bingo!

When I woke, I was convinced that trust in the truth of a statement is a cerebral process that can happen independently of any rigorous appraisal of that statement, as if faith in something could precede consideration of the thing's actual content. Along with this shocking realization, I concluded from this experiment that marijuana was *not* a great factory churning out brilliant ideas but rather—and only—a factory producing interesting ideas in the internal world, ideas that at the moment of their conception can seem wonderful, but whose successful use in the external world requires a critical evaluation a posteriori, after the marijuana's effects have come to an end. I started to talk about my state of consciousness under the influence of marijuana as "the other," marking my first conscious experience of an alterity, a parallax, a difference in internal perspectives that created the possibility of a dialog within my own mind. I and I.[2]

And from out of this internal dialogue came a greater tolerance toward external dialogue. In the words of Norman Mailer,

One's condition on marijuana is always existential. One can feel the importance of each moment and how it is changing one. One feels one's being, one becomes aware of the enormous apparatus of nothingness—the hum of a hi-fi set, the emptiness of a pointless interruption, one becomes aware of the war between each of us, how the nothingness in each of us seeks to attack the being of others, how our being in turn is attacked by the nothingness in others.[3]

The increase in sociability is a notable effect of marijuana, as described by John Lennon:

Marijuana was the main thing that promoted non-violence amongst the youth, because as soon as they have it, they—first of all you have to laugh on your first experiences. There's nothing else you do but laugh, and then, when you've got over that and you realize that people aren't laughing at you, but with you, it's a community thing, and nothing would ever stop it, nothing on earth is going to stop it, and the only thing to do is to find out how to use it for good, or for the best.[4]

While my travels proceeded without my seeking out any more experiences of that kind, the encounter with the weed on the island of Chiloé was a professional and personal watershed for me. First, for its prompting me to send a letter to my brother suggesting we meet somewhere halfway for a reconciliation. The meeting happened, we recreated our state of harmony, and there we remain to this day—I and he, he and I. Second, for its allowing me to see with clarity and enthusiasm the research subject to which I would thereafter devote myself: the brain.

After my return from that trip around South America, there followed a few years in which the flowers came intermittently onto my radar, but I never sought them out. When we met, I would enjoy the state into which they'd cast me, but I was desperate to do something under their effects that was not just super-relaxed chatting at a party. I graduated, completed my master's, and was well into my doctorate when things began to change.

The starting point for that change was a romantic disappointment that shook my previously happy graduate journey. After getting unceremoniously dumped by telephone, I went into a spiral of suffering that lasted several months. Then one fine day, as I was lamenting my fortunes on an international call with a dear old friend, the historian Cristiana Schettini Pereira, I received a very valuable piece of advice: "Kid, you got to smoke more marijuana." All tangled up in my pitiful rejection pains, I was persuaded by my wise friend's practical advice. The increase in frequency of my chance encounters with the weed soon helped me to stop mulling over memories of my thwarted love.

Still, I did resist having the flowers at home, out of a series of prejudices, fears, and taboos. I learned to seek them out through a slow process of approach. In this process of gradual familiarization, I was able to benefit from the generous guidance and patient welcome of two beloved friends, a brilliant couple, in science and the arts, who are very expert in knowing how to live with ganja, managing as they do to maintain a steady combination of high professional achievement and fun cannabis recreation, in a life full of picnics and dinners, movies and books, exhibitions and concerts, walks and travels. Bit by bit, my friends taught me to increase the possibilities for action and reflection under the flowers' influence—no stress, all chilled, all cool.

When it came time for me to write up my doctoral thesis, I locked myself at home for a month with large doses of Mestre João Grande, Gilberto Gil, Gal, Paulinho da Viola, Janis Joplin, Clube da Esquina, Maria Bethânia, Van Morrison, and Pink Floyd. My diet was rounded off with miso soup, chocolate biscuits, a lot of coffee, and a lot of ganja. For days and nights with no beacons to mark the time, I would wake up and fall asleep wrapped in a cloud of thoughts that spun around in search of expression. At the end, totally lucid, smoking nothing, I edited the entire document until any scaffolding had been removed and left no trace. It worked out.

Though I had started my doctorate six months late and had really struggled to adapt to winter in New York and all that city's challenges, after five years and under the careful supervision of my advisers I managed to make a number of scientific discoveries about the expression of genes that regulate learning in songbirds and dreaming rats. This success allowed me to be accepted for a postdoc at Duke University.

It was also what got me the invitation from Rockefeller University to be the class-of-2000 valedictorian. In a suit and tie for the first time in my life, I gave my speech, watched by my proudly beaming mother. On our way home, arm in arm with her, I realized that I'd found the best moment to come out of the closet in relation to the flowers. I understood that this was the perfect opportunity for healing in our family. After all, no one could say I was messing up my life. I was very aware that the role of the good kid that I was playing with increasing hypocrisy up until that moment had contributed quite a lot to the stigmatizing of my brother within the family. It was time to correct the mistake that blamed those flowers for the problems in people's relationships.

I can't say it was easy. For days I kept postponing the conversation and suppressed my desire to light up, waiting for just the right moment to appear. Things began to accelerate when my aunt Vilma arrived for us to take a celebration trip to Quebec. When we left New York, I had a joint in my pocket and an idea in my head. At one point, like it was no particular big deal, I asked my mom what she'd do if she had a habit that she found very pleasurable but which displeased other people. She said she'd keep doing it, without a moment's hesitation, it was a matter of autonomy and freedom. Next, without a moment's hesitation, I lit the joint.

Mom was shocked. She really hadn't been expecting that! She went berserk, got furious, really enraged. Then she put on a sulky pout the size of the world and gave me the silent treatment for three long days. Curiously, it was my aunt, an incredibly observant Catholic, a devotee of St. Expeditus, who started to pacify my mom's rage. After all, it wasn't like Mommy's little darling had suddenly turned into a whole different person just because of this, right?

No, I was the same person I'd been before, albeit a little

more honest and real. By the time we returned to New York, I could already light up in front of my mom without her suffering too much. In time, her prejudice dwindled, died, and became a compost for congeniality. There began a slow but inexorable process of restored closeness between her and my brother, who gradually was integrated back into the family where he belonged. In the decades that followed, Júlio graduated and completed his master's in architecture and town planning at UNB, specialized in airport design in a sandwich doctorate at Pennsylvania State University, and learned to paraglide.

The repositioning of marijuana within our family context gave me new horizons. From the year 2000, the flowers began to be a part of my routine just like coffee and sugar. In time—glory be!—I managed to quit sugar almost entirely. Those flowers, on the other hand . . . I started venturing into the universe of improvising under their influence, in jam sessions of verse and prose and rhyme, on the atabaque, the berimbau, and the agogô, both with musician friends and in the free and flexible practice of the movements of Capoeira, which I'd started learning in 1999.

Very often, while writing this book, I have found myself acutely aware that I was dealing with something that, just like Capoeira, had been banned, persecuted, and demonized; associated with the Africans who were brought to Brazil enslaved; and accused of promoting insolence and being damaging to society—and which today, in spite of it all, is loved across the planet.

My keeping up this habit and the expansion of my capacities to carry out different activities under the influence of the flowers made me understand that they have the special gift of transforming any situation into something even more interesting, except when the reality principle outweighs the pleasure principle. Stress, fear, competitiveness, and the need for high-precision performance tend not to mix well with flowers that are rich in THC. When we are responsible for somebody particularly vulnerable, such as a baby or someone who is convalescent, it's sensible to avoid them. Flowers rich in CBD, meanwhile, might help or not, but they don't tend to get in the way except for the sleepiness that they can induce.

But at least when we're talking about a situation that doesn't entail major risks, these flowers are the promise of a fun adventure. Many people work better under their influence, a capacity that one acquires with practice, and which can be related to hyperfocus, to creativity, to a drop in anxiety, and the capacity to go harmoniously with the flow of one's surroundings. Other people, however, whether because they have certain genetics or just because they are still relatively inexperienced with the weed, cannot use it to produce anything worthwhile in a timely manner. Different strokes . . .

The scientific research that has been carried out on cannabis use confirms that ingesting the weed prompts more favorable assessments of the creativity of one's own ideas and those of others,[5] but it also shows that high doses of THC can be genuinely harmful for creativity.[6] A piece of research carried out on 721 participants showed a greater perception of creativity and better performance in a creative task among moderate cannabis users than in nonusers, possibly because they have a greater degree of openness to the experience.[7] Another study, on 160 participants, showed that marijuana was able to increase the verbal fluency in people who are not very creative, but not in people who *are* very creative.[8] Similar results were obtained from a subsequent study of 148 participants that also confirmed an increase in creativity only in low-creativity individuals, which also suggests a ceiling effect.[9] Research carried out on the psychological profile of marijuana users, involving 278 university students, showed a link between marijuana use and higher scores for creativity, a spirit of adventure, and a search for novelty in internal feelings, as well as lower scores for authoritarianism. Curiously, after two years of use, boredom with the environment was reduced significantly.[10]

Boring situations cry out for marijuana. You can take your pick: waiting in line at the bank, a movie that's not great, some repetitive chore, a white wall with a fly buzzing around . . . All of these entirely dispensable parts of everyday life can become a source of actual satisfaction when you're able to bloom. For someone who likes it—which isn't everyone, of course—any situation, no matter how banal, is transformed into an opportunity for reflections and insights that are existential, anthropological, comic, aesthetic, sensual, spiritual, or whatever you feel or want. In the words of the writer Kurt Vonnegut, marijuana makes "stress and boredom infinitely more bearable."[11]

In many different situations, marijuana is a balm for our existence, as it significantly improves the flavor of dull experiences through sensorial alterations, changes in the perception of time, and intense physical and mental relaxation. Meanwhile, it doesn't often compromise experiences that are pleasurable. On the contrary, even the best experiences can be improved with the right flowers, in the right quantity, at the right moment. But it's always necessary to emphasize that not every experience lends itself to having the flowers around, as we have seen above and as we will see again.

It's important to remember that what exists is not marijuana but rather marijuana*s* in the plural, so extensive are their genetic possibilities and the techniques for extracting their active principles. In addition, the genetic variations in the people using the flowers and the social context of that use determine a good part of what might happen in each case. Which is precisely why, in the process of rediscovering cannabis therapies that we are currently going through, the cumulative knowledge of frequent users is often some way ahead of what science has thus far been able to describe using rigorous clinical tests. This happens because biomedical research suffers from a terrible inertia caused by its conflict of interest with the pharmaceutical industry, who sees no benefit to spending millions of dollars to research a plant on which they cannot have a monopoly. In epilepsy, in autism, in depression, in Parkinson's or Alzheimer's diseases, and in so many others, it is the qualitative experimentation, carried out in real life by children, adolescents, and adults, that's the light source pointing the way for quantitative scientific research, as regards the fundamental biological mechanisms as well as in the clinical setting.

The fight to legalize cannabis therapies, involving as it does so many legitimate stakeholders, broadened its scope well beyond medicine, reaching nursing, speech therapy, psychology, physiotherapy, and physical education. Inevitably, we must also consider professional veterinarians, who typically have more freedom than human-health professionals to learn in practice about what does or doesn't work.

Since 2018, the veterinarian Tarcísio Barreto Filho has been treating dogs and cats with marijuana oil.[12] His first patient was Prog, a six-year-old poodle with epilepsy that resisted the effects of other kinds of medication. Tarcísio says that "after years of using potassium bromide and pheno-barbital, the animal was bloated and presenting with constant mood swings, very sad one moment, agitated the next. Before cannabis, he was having fits twice a week; after cannabis, the fits started to happen once every thirty or forty days." From 2018 to 2023, Tarcísio treated around seven hundred pets with cannabis oil. He reports that in general "the animals were calmer, more focused at play, with their digestive systems well regulated, good physical posture, and more balanced behavior." The cannabis vet has had more and more scientific backing, since various studies have now confirmed the usefulness of marijuana in treating anxiety, inflammation, and pain in nonhuman animals.[13] It's natural that this should be so, since the endocannabinoid system is present in all vertebrate animals.

Fortunately, this long list of therapeutic revolutionaries focusing on cannabis also includes those patients and relatives involved in cannabis therapies, who often establish relationships with their health teams that are very horizontal and encourage more autonomy than conventional medicine tends to allow. Therapeutic cannabis is a joint health care effort that benefits from group participation.

Loving the Flowers Too Much

Anything in excess can cause issues, and it's no different with marijuana flowers. Any and every substance has its risk groups, people for whom, whether down to their genetics or their lifetime experience, that substance can cause problems even in small doses. For one instructive example, consider the consumption of dairy products. People with low levels of production of the enzyme lactase cannot digest lactose and so have an intolerance toward animal milk and its derivatives. It is estimated that this condition affects between 57 percent and 65 percent of the global population, with higher prevalence among those who are not white.[1] Another useful example might be those people with particular variations in the genes that codify the aldehyde dehydrogenase and alcohol dehydrogenase enzymes, who struggle to metabolize ethanol and are thus hypersensitive to alcoholic drinks.[2] These genetic variations are more prevalent in certain East Asian populations, such as in China and Korea, but rarer in others.[3]

While we might all agree that it is necessary to protect people who are vulnerable to lactose, no one is suggesting that we should be banning milk in order to protect them.

Same goes for ethanol, with the exception of those Muslim cultures where it is banned outright. Almost everywhere in the world, protection for people with an intolerance to milk or alcohol comes from the reliable information found on product packaging and the existence of alternatives on the market, like lactose-free cow's milk or plant milks, and non-alcoholic beers and wines.

Biomedical research shows that the risk groups for marijuana are: 1) pregnant and lactating women; 2) children and adolescents; 3) people with a genetic propensity toward psychosis; and 4) people with depression. Group 1 has been receiving growing support from research showing risks to the healthy development of the fetus and the baby.[4] Group 2 has ample support in the scientific literature already, since early marijuana abuse is associated with an increased risk of amotivational syndrome, characterized by apathy, passivity, and an aversion to goal-oriented behaviors and harm to academic performance.[5] While it might prove benign in isolated cases, on a populational level, the consumption of marijuana is contraindicated for young people who are healthy and do not need it as a medicine for some specific condition. A brain imaging study involving 799 participants recently suggested that cannabis consumption between the ages of fourteen and nineteen can be connected to an excessive reduction in the thickness of the prefrontal cortex.[6] A piece of advice for parents and their children: The consumption of marijuana should be delayed as far as possible, since the adolescence that concludes legally at the age of eighteen extends in the brain to about twenty-five, when the circuits of the prefrontal cortex that we need for making decisions unimpulsively are fully developed. Group 3 relates specifically to THC and not to CBD, as we saw in the last chapter, "Maturana, Marijuana, and the Green Frog." Group 4 is the most disputed, as in low doses, marijuana tends to have antidepressant effects, while in high doses it can be associated with amotivational syndrome and be pro-depressant.[7]

Beyond cannabis's clearly delineated risk groups, consumption in excess can be harmful to anybody. I've known people who have allowed their entire home, their leisure time, and their work to be overtaken by the whims of their dog. Just as one's day-to-day life can become stressful if a dog comes to occupy too much space in the life of their owner, so allowing marijuana to expand its place until it is occupying every waking hour can lead to a range of problems, such as vagueness, inefficiency, laziness, and insomnia, as well as the mouth and lung problems arising from smoking it and possible cardiovascular trouble.[8] In some people, the excessive use of cannabis can lead to complex disorders, such as hyperemesis, a rare condition involving vomiting and recurring nausea.[9]

It's important to know that there are many healthier ways of consuming cannabinoids than the burning of a marijuana joint, such as vaporization for inhaling, oils for absorbing sublingually, and many other formulations, from creams and cookies to candies and suppositories. Still, it's worth mentioning that even the consumption of marijuana by smoking is not accompanied by the high cancer risks that accompany smoking tobacco, possibly due to a compensatory effect produced by cannabis's antitumor properties.[10]

In a sense, marijuana is like chocolate, french fries, or

popcorn: For somebody with the genetics and the personal history to like it, the more they consume, the more they will want to consume. What are the consequences of an excessive dose of marijuana? For most people, just a lot of sleepiness. To die from marijuana, you'd need someone to drop a large brick made of compressed marijuana onto your head. Other damage can be done, though.

It's estimated that marijuana leads around 9 percent of its adult users into dependency—for comparison, tobacco causes dependency in around 67 percent of those who use it. One of the great dangers in excessive consumption of the flowers is allowing time to pass without doing anything worthwhile, like social-media-scrolling marathons for hours on end. And with frequent repetition of this ingestion comes tolerance, a loss of the pleasurable sensation, and an increase in the dosage and in lethargy. Paranoia and uncontrolled appetite can also become obstacles to someone who loves these flowers too much, especially if the THC-to-CBD proportion is very high. Also, for most people, the consumption of marijuana shortly before sleeping has a clear negative impact on one's dreams. It doesn't do away with them entirely, but it does significantly damage the capacity to recall them, turning the memories fleeting. We still don't know why this happens, nor which cannabinoids are responsible for this effect. Some old studies show that THC causes a reduction in the amount of REM sleep, during which we have our most vivid and complex dreams, but we know nothing about the plant's other hundreds of molecules. In order to protect your dream experience, avoid consuming marijuana at night.

Another aspect that we should consider is the effect of

marijuana when driving a vehicle. Consuming the flowers does tend to increase reaction times to sensorial stimuli, but then it does also tend to reduce aggressiveness and speed in traffic.[11] This compensatory effect might explain why marijuana, unlike alcohol, shows no dose-dependent relationship with traffic accidents, except at very high doses.[12]

Sagan described his own experience thus:

> I have on a few occasions been forced to drive in heavy traffic when high. I've negotiated it with no difficulty at all, though I did have some thoughts about the marvelous cherry-red color of traffic lights. [. . .] I don't advocate driving when high on cannabis, but I can tell you from personal experience that it certainly can be done.[13]

Once again, we must keep in mind the huge variety in the genetics of the plants and of people in order to understand why some drivers experience hyperfocus, pleasure, and security when driving under the influence of cannabis while others feel uncomfortable or even unable to do it. Scientific research has shown that cannabis is much safer for driving than alcohol[14] but also that a high dose of THC can be harmful for driving up to a hundred minutes after consumption, while one's capacity to drive remains intact with CBD even in high doses.[15] In addition, driving under the influence of marijuana can be worsened in situations in which one's attention is divided or in those of increased complexity.[16]

As for the regulating of driving under the effects of cannabis, perhaps it's more important to consider what the

THC-to-CBD proportion is than merely estimating the THC concentration, but the question remains contentious and demands further research.[17] Meanwhile, there is no longer any doubt that the consumption of marijuana and alcohol in combination has adverse effects that are magnified, seriously damaging a person's ability to drive.[18]

Banning the Flowers

Beyond all the real or perceived risks associated with marijuana consumption, we need to discuss the biggest of all the dangers caused by a love of these flowers: The fact that they are banned, which creates risks not only for users but also for all those around them. Unfortunately, a lot of relatives of patients who need marijuana oil are still planting and harvesting their flowers illegally. To make it worse, while therapeutic marijuana has been increasingly emancipated by the majority-white middle classes, many mothers and grandmothers—almost always Black and poor—continue to lose their children and grandchildren to the war on drugs, which is the main cause of incarceration and death among young Black people in Brazil.[1] This daily massacre, authorized by the silent majority, has already mown down countless children like Ágatha Félix, struck in the back at eight years old by a bullet fired by the military police.[2] Ágatha was inside a VW camper, sitting beside her mother, on her way to school in Rio de Janeiro's Complexo do Alemão favelas.

In the words of Renato Filev, neuroscientist and anti-prohibitionist activist:

It's impossible to separate therapeutic use from
the political and regulatory implications around
marijuana and its social use. Patients who cultivate it are prosecuted and incarcerated for
growing their own treatments. Firms distribute,
sell and profit from the same product that jus-
tifies the killing of people who might or not be
connected to its unauthorized dealing. The ap-
parent policy of public safety supported by the
criminalization of and fight against marijuana
and its derivatives should be stopped at once,
and policies of reparation implemented to reduce the vulnerabilities and the needs of those
people and communities affected.[3]

The war on drugs is much more toxic than any substance
and compromises all three dimensions of use: substance, set
(the body), and setting (the context of use). For the substance,
banning prevents people from knowing dosages, expiry dates,
and whether there are contaminants. For the body who receives the substance, banning makes it impossible to have a
discussion about risk groups and problematic use. For the social context, banning creates toxicity and paranoia, through
violence and marginalization.

I'm sometimes asked if I'm in favor of freeing up access to drugs. The truth is, a general liberalization already
exists nowadays, since anybody of any age can get hold of
any drug—though they have no guarantee that they are in
fact buying what they think they're buying. While certain
drugs like marijuana are banned and demonized, others like

alcohol are glorified openly on television, with no scientific standards implemented nor any protections for society, while stray bullets continue to fly over our heads.

The rapper Mano Brown points directly at the racist and classist contradictions inherent in the war on marijuana users:

> The rich smoke, the poor smoke, Blacks smoke, whites smoke, everybody smokes [. . .] how are you going to distinguish between those young people, there, amid all those problems, so many traumas, often social traumas, of the collective, which relate to the life of the individual, a guy is carrying the burden of those obligations, and goals not reached, and there's a lot that's blamed on the consumption, and that's the thing, sometimes the guy's only company is a joint, where he can stop and just think for a bit, sometimes he hasn't got that friend, he hasn't got a dad, an elder figure, and he's got, I dunno, in that one drag, oh man, he's got a brief lull in that awareness that things aren't too easy . . .[4]

Brazil's current Drugs Law (law no. 11.343/06) is directly linked to over-incarceration, in that it increases the penalty for dealers and reduces it for users but without defining objective criteria to separate the two.[5] That decision comes down to the individual judge, and therefore also to the police authority who decides in the heat of the moment which real-life situations ought to become a recorded crime, a shadowy bribe, or a one-way trip to hell. A Black person arrested in a favela with a few grams of marijuana has a high chance of being locked up as a dealer, while a white person found in a wealthy neighborhood with large quantities of the same drug is likely to be categorized as a user.

In other words, the subjectivity of the distinction between users and dealers legitimizes biases of class, race, and gender within the state authorities. This results in a vertiginous increase in the prison population, the majority of whom are serving sentences for drug trafficking. In fifteen years, we have gone from 240,000 people in prison to over 800,000. Today more than 60 percent of incarcerated women are in for dealing, and they are often caught when bringing drugs to their partners in prisons. Often they never go back to their children. They are given long sentences that only serve to destabilize their families still further.[6]

To complete the Dantesque picture described above, any prison officer will tell you what happens when there is a shortage of marijuana in prisons: "The jail gets turned upside down"—that is, rebellion breaks out. The terrible irony is that without marijuana to help the inmates keep calm, the state quickly loses control of the penal system that is enforcing punishments for marijuana possession among the most vulnerable.

This thorny subject, which prohibitionists treat as a taboo, is a pitiful open secret, which no one will admit to knowing even though everybody knows. Since nobody wants to talk about it, I asked AI—which in theory is a robot apart from any possible explicit pro-marijuana biases—what happens in a prison when there is a lack of marijuana for the inmates. This is ChatGPT's response:

1. Increased tension and violence: The lack of marijuana or other drugs can cause frustration and irritability among those inmates who use them. This can lead to an increase in tension and potentially to conflict between those inmates who use them.

2. Search for substitutes: In some situations, inmates can look for alternative substances to replace marijuana, such as tobacco, homemade alcohol, or even harder drugs. This can lead to health problems and high-risk behaviors within the prison.

3. Increase in internal dealing: The shortage of marijuana can lead to an increase in internal drug-dealing, with the inmates themselves attempting to meet the demand and to profit from the sale of illicit drugs within the prison. This can lead to an increase in violence and instability within the penitentiary setting.

4. Impact on the prison's internal economy and hierarchy: Marijuana and other drugs very often have an economic value within the prison system. The lack of marijuana can affect the informal economy and the power dynamics and hierarchy between the inmates.

The anti-prohibitionist lawyer Cristiano Maronna sums the crisis up well:

Changing the direction of drugs policy is an urgent matter, because this punitive and repressive approach has failed. Besides illegal drugs circulating almost entirely freely, despite being prohibited—which is the paradox of prohibitionism—we have seen effects that are more harmful than drug abuse itself. The war on drugs produces corruption, violence, over-incarceration, and a strengthening of criminal organizations.[7]

Ricardo Nemer, another anti-prohibitionist lawyer, can't hold back:

It's really hard to keep one's romantic views intact knowing that these people who have never planted a flower, or who have actively resisted the sustaining of that tradition, are going to be the ones to dominate the legalized market [. . .]. The business community is going to keep on profiting from everything, whether prohibition or legalization. They've never had their homes strafed with machine-gun fire, blown up or raided by the police. [. . .] Activists from the Marcha das Favelas, for example, were tortured by officers from the Bope squads, and informally banned by means of torture and threats carried out by the police. Today militants aren't allowed to do their march within the favelas. We know that in the properly paved areas of the city, people get treated differently, white folks from the middle-class smoke marijuana without having their rights violated or suffering any consequences. The only flowers that the favela youth will get are going to come in wreaths and they'll continue to be remembered only in graffiti on the walls, portraits on T-shirts bearing the words "Miss You Forever" or a smiling photo on their mother's bookshelf.

[. . .] The *entressafra* is the inter-harvest season, a period between one cycle and the next, and it's an interval characterized by the shortage of crops available for immediate consumption. It's the in-between time where the grower is preparing the earth for the planting of the next crop. [. . .] Society and the fable of equality and fraternity have been promoting a shop window for consumers and absolutely no opportunities for those young people who see drug dealing as a possibility for survival and a way out of invisibility, of subalternity, and of work precarity. If a rich person is smart, a poor person is a rogue. [. . .] The term "cria de favela" refers to someone who comes from a historically stigmatized social setting, with low levels of development owing to a lack of public policies and opportunities for social mobility.

[. . .] The dark *entressafra* of the ban, prompted by the war on drugs, was just one more cycle of "legal *entressafra*" in Brazil. I could give you a list

of the children killed with firearms in the met-
ropolitan area of Rio de Janeiro who died this
year and the list keeps on growing. Of the regular folks shot down on their way to church. The
thousands of people who are scared, even in
their own homes, because of the war on drugs,
the relatives of the victims of the violence of the
war on drugs. But I'm also going to talk about
the workers in the favelas: the informal ones, the
formal ones, and the illegal ones. [. . .] A dealer
is a worker whose work has been made precar-
ious. Their work is regular and the worker has
a boss, receives orders, and a wage, and if they
make a mistake, they aren't let go, they pay for
it with their actual life. The criminal gang is one
of the structures of a complex political system
of soldiers, police, and public authorities for sustaining the production of violence and political
capital. Just think, who is it that profits from the
ban? If today we know more and have access to
marijuana in Brazil, that's down to the young res-
idents of these places. It was the favelas, without
them we wouldn't even have marijuana. They
stored it, they distributed it, they resisted. They
literally sustained the tradition and resistance to
the "*entressafra* of the ban," paying for it literally
with their own lives. How are we supposed to
think about marijuana boutiques when there's
still a war? We need to make the government,

businesspeople, businesses, and politicians really commit to this conversation. Society needs
to be talking about ending the war and building
a peace agreement and transitional justice. The
only people afraid of this pact, with an end to the
violence, an end to the deaths of poor people,
are those profiting from the ban: the networks of
police and military officers working protection
for kickbacks at the locations where drug sales
happen. As police chief Orlando Zaconne put
it very well, drug dealers are the "shareholders
in nothing,"[8] as they only manage the shop for a
small group of powerful people, they're just front
men, cannon fodder, people who are "dispos-
able" and "killable." The more war there is, and
the more fights and factions, the more ammu-
nition and weaponry will be sold, and the more
innocents will die. We need to be creative and do
things differently to how we've done them in the
past. Why should a pupil from a poor family in
public school deserve less state investment than a
poor person who's imprisoned?[9]

I think about Ágatha Félix and about her mother, I think
about my own children, and I cry. In such hours of despair,
faced with this horror, I seek refuge in my profession of faith.
I am consoled by the lucidity that comes from knowledge.
At their 2018 annual meeting, the Brazilian Society for the
Advancement of Science (SBPC) unanimously approved a

motion that proposed that all drugs should be legalized and properly regulated depending on their specific benefits and harms, so that we might thereby protect risk groups and respect users.[10] No substance is simply divine or demonic. We need a legalization that's fair and equitable, stripped of all prejudices, based on science and on love for our neighbors.

If we want to carry out a rigorous appraisal of the psychoaffective effects of marijuana, we need to be able to quantify the extent to which the war on drugs is *itself* toxic, producing violence, fear, paranoia, and anxiety. As we're reminded by João Menezes, a neuroscientist who is an activist for marijuana legalization, this is a debate in which "we need to take the fear out of our hearts."

One recent piece of research that was rigorous and well controlled shows that the complete legalization of marijuana, both for medicinal use and for recreational use, is a significant step on the road toward social healing. The study looked at 4,043 twins living in the United States in regions with different marijuana regulation policies.[11] The participants were evaluated in adolescence and in adulthood, which made it possible to test out the effect of legalization over time and to evaluate interactions with factors relating to possible vulnerabilities, such as age, sex, and mental disorders.

The study showed that those individuals who lived in states where recreational use is legalized consumed cannabis more frequently and had fewer problems with alcohol than their twins who lived in prohibitionist states. The legalization of marijuana was not associated with any adverse results on a populational level, including its problematic use or a predisposition for certain mental disorders. This powerful study is an anvil dropped onto the argument that marijuana is a factory of familial and social maladjustment. Which is not to deny that for people with certain vulnerabilities, marijuana can be dangerous, but rather to state that if people can take the fear from out of their hearts, these flowers will smile and can do almost everybody some good.

Getting Old with the Flowers

With the exception of children and adolescents with some particular pathology, marijuana is for older people. Since everybody with any luck grows old, it's worth heeding the words of Felipe Faria, the president of the Reconstruir patients' association: "Don't wait till you need it to be in favor of it."

For the last three years, I have been treating my anxiety and joint pain with cannabis oil, on a medical prescription. Encouraged by my partner, neuroscientist Luiza Mugnol-Ugarte, and with the advocacy support of lawyer Marina Bortolon Moreira, I requested and obtained a preventive habeas corpus in 2021 from the federal district court of Rio Grande do Norte to be able to plant marijuana, carry flowers, and produce my own therapeutic oil. The judgment was overturned in the secondary courts, but we joined forces with Rede Reforma to take the request to the STJ. Finally, in July 2023, we were granted a favorable decision in the higher court. Freedom, albeit belated.

Getting old without unbearable pains in one's body and mind is a blessing that all people deserve. Chronic pain is one of the main indicators for people to use therapeutic cannabis.[1] Biomedical research on THC-based or CBD-based products show that both produce relief when it comes to chronic pain,[2] reducing the intensity of the painful feeling and its interference and improving the patient's quality of life, general health, mood, and sleep.[3] We're not talking about a conventional analgesic effect here, which blocks the feeling of pain in a peripheral nerve or at the level of the central nervous system, but rather of a powerful distracting effect, which pulls focus from the pain and integrates it into a landscape of perceptions and thoughts that is far larger and more bearable. An examination of the amazing case of a woman insensitive to pain, with extremely low anxiety levels and rapid scarring of wounds, revealed a genetic mutation that reduces the levels of the enzyme responsible for the degradation of anandamide.[4] The research points toward the direct involvement of the endocannabinoid in pain relief.

The broad range of uses of marijuana for relieving the unwanted consequences of senescence constitute a profound popular knowledge, which insists on sprouting human life through the gaps in the cracked concrete of prohibition. Consider what happened in Cruzeta, a small town of eight thousand people located 230 kilometers from Natal (RN). In the 1980s, one elderly resident began growing marijuana for therapeutic purposes, as if it were just lavender, gorse, or any other of the many medicinal plants used in bottled form, in tinctures, or in infusions. In time, his neighbors started asking for seedlings of the plant that they called *liamba*, which had gained a reputation as a miraculous treatment for pains, epilepsy, respiratory problems, cancer, migraine, and other ills. In 1996, an anonymous tip-off mobilized the police, who discovered enormous plants in several homes in Cruzeta, as well as in the cemetery, in the town's leafy squares, and in front of a church. The case got national attention and the town's inhabitants became quite alarmed at the prospect that they might be arrested for planting marijuana. However, because they consumed its flowers and leaves only in the form of tea, they ended up not being charged. Still, all the plants were pulled up by the police, the crops were incinerated, and the elderly people were not only deprived of their medicine but were obliged to attend deterrence courses on the dangers of marijuana.[5]

Now that the therapeutic use of cannabis is beginning to be accepted in a large part of the planet, we mustn't forget that for the purposes of effective therapy, there is not one single marijuana that exists but many different marijuanas. Beyond all the plant's own genetic complexity, its effects always—crucially—depend on the genetics of the person using the flower, as well as on their personal background and the social context of its use. When there is high-quality information about the concentration levels of many different molecules, from cannabinoids to terpenes to flavonoids, marijuana lends itself well to medicine that is personalized, suited both to the individual, at their specific moment in life, and to their pathology, at its particular stage of development. For each patient, at each moment, there is a masterly formula in the shape of this flower that can be sought, found, and cultivated.[6]

At Israel's Technion Institute, David Meiri and his team of researchers use robots to measure and combine precise quantities of substances in order to carry out a very controlled investigation of the therapeutic efficacy of multiple combinations of cannabinoids in cell cultures derived from tissues extracted from the patients themselves. This "bottom-up" strategy seeks a personalized formula for the treatment of each type of cancer, at each stage of the illness, in each individual.[7] In the Netherlands, corporate scientists are working on the reverse path, "top-down," looking for the whole flower that in all its chemical complexity is the best suited to each patient.[8]

Over the decades of life shared with their flower-using children, my mother Vera and stepfather Edson gradually made their peace with marijuana. In time, at our big family Sunday lunches, the family's favorite drugs—alcohol, sugar, and animal fat—have come to be joined with barely a squeak by the flowers enjoyed on the veranda. But this tolerance for the cannabis habits of the next generation had not turned into curiosity or any desire to try it on the part of the older one. Each to his own.

However, at the end of her life, afflicted by depression and fibromyalgia, my mother found in these flowers an unexpected respite. The turn came one Christmas Eve, as the house was buzzing with relatives and friends, but she refused to come out of her room. The big dinner was ready to be served and there was still no sign. I went to fetch her and found her prostrate on the bed, in terrible pain and a terrible mood. After some insistence on my part, she agreed to smoke a joint. She inhaled, coughed a little, her mind wandered for a few moments, then finally she went off to have her bath. When she reappeared, she had that youthfully happy expression we all so loved in her. She came downstairs to the crowded room and the party began. We shared, sang, danced, and had one more happy night.

Dying and Being Reborn with the Flowers

And then the day comes when it is time to die. Faithful as any dog, marijuana is an excellent plant to accompany you in those last moments before the final goodbye, lending itself well to the palliative care of the pains of the person languishing,[1] as well as the anxiety surrounding the imminent death.[2] In the words of Dr. B. J. Miller, marijuana

> can help some of our patients not only cope with suffering but also reframe their perspectives. [. . .]
>
> We should be keen, as palliative care specialists, to look for ways in which we can help make our patients' lives *more* wonderful, not just *less* terrible. [. . .]
>
> To these good ends, marijuana can be one helpful tool. By wielding rather than excusing its psychoactive properties, we can alter our points of view. The human penchant for perspective-making is one of our sharpest tools and most certainly under-utilized in the clinical setting. Work with your patients on *how* they see things—not

just on *what* they see. Often enough, a diagnosis
is, among other things, an invitation to revisit how
we see ourselves and the world we're still part of,
or to think again about how we're living, or to
strike a playful tone with reality. Not simply thinking our way forward, as an intellect, but feeling
our way forward too.[3]

In his final interview, which he gave just a few days before he was taken away by a devastating cancer, the journalist
and writer Otavio Frias Filho bore witness to the inestimable
value of marijuana for terminal patients: In reducing their
anxiety, encouraging their imagination, and increasing the
sense of flowing through time effortlessly, marijuana helps
many people to deal with the fundamental anxiety about
the inevitability of their own death, the universal beacon of
human consciousness, the primitive source of all pains and
fears.

It's worth reading the testimony of Melissa Etheridge,
two-time Grammy-winning singer-songwriter, on the role
that marijuana played in her treatment for breast cancer. "Instead of taking five or six of the prescriptions, I decided to go
a natural route and smoke marijuana," she said.[4]

So when I was dealing with cancer, I realized
that so much of it is what we manifest in our
thoughts. I mean, every religion has this in it.
Every spiritual practice understands this. And
it's just us getting back to that—and cannabis
helps us.[5]

If we are to deal with death, we must renew the art of living. We can only do it by changing, adapting, and continuing to change more, for as long as possible. Before I was familiar with marijuana, I was not more studious, more committed, or more disciplined than I am today. What I was, certainly, was more arrogant, boring, competitive, and inflexible. And less creative.

Of all the great gifts that these flowers can grant you, cognitive flexibility is one of the more benign and transformative, both in itself and in relation to sensorial stimuli, to people, animals, plants, and inanimate objects. And—such a blessing—another great gift usually accompanies the first: free contact with one's emotions, which makes it possible to have a deep relationship with beauty. Listening to music or watching a movie under the effects of marijuana, finding it easier to access the tears and laughter that purge the soul.

Sagan saw these uses clearly:

> The cannabis experience has greatly improved my appreciation for art, a subject which I had never much appreciated before. The understanding of the intent of the artist which I can achieve when high sometimes carries over to when I'm down. This is one of many human frontiers which cannabis has helped me traverse. [. . .]
>
> For the first time I have been able to hear the separate parts of a three-part harmony and the richness of the counterpoint. I have since discovered that professional musicians can quite easily

keep many separate parts going simultaneously
in their heads, but this was the first time for me.
Again, the learning experience when high has
at least to some extent carried over when I'm
down. [. . .]

Cannabis enables nonmusicians to know a
little about what it is like to be a musician, and
nonartists to grasp the joys of art.[6]

In the words of Gilberto Gil, famous flower lover: "Marijuana helped me with my music, I always say that without
any doubt at all. Marijuana helped me with creativity, by the
way I used it. For the kind of use I wanted to make of it, oh
yes, it did help me."[7] From Louis Armstrong to Bob Marley,
from Rita Lee to Madonna, from Gal Costa to Lady Gaga,
from Snoop Dogg to Jards Macalé, from Fela Kuti to Tupac
Shakur, from Hélio Oiticica to Antonio Peticov, from Maya
Angelou to Kurt Vonnegut, from Allen Ginsberg to Waly Sa-
lomão, from Bob Dylan to Marcelo D2, from Cássia Eller to
Mano Brown, from Zé Celso to Maria Alice Vergueiro, from
Júlio Bressane to Renato Russo, from João Gilberto to Keith
Richards, from Bia Lessa to Glauber Rocha, from Rihanna
to Jimmy Cliff, from Miles Davis to Janis Joplin, from Peter
Tosh to Tim Maia, from Gregório Duvivier to Morgan Free-
man, from George Harrison to John Lennon, from Bezerra
da Silva to Chico Science, from Céu to Sagan, marijuana
fertilizes creativity and helps life to flourish for as long as one
is able to live.

Reconnecting with the simple pleasures in life—the art
of living—is the cannabis gift that is so well described by

the poet Maya Angelou: "Smoking grass eased the strain for me. [. . .] From a natural stiffness I melted into a grinning tolerance. Walking on the streets became high adventure, eating my mother's huge dinners an opulent entertainment, and playing with my son was side-cracking hilarity. For the first time, life amused me."[8]

When you hear Louis Armstrong singing the most famous song he ever recorded, enumerating all those evocative things that make the world so wonderful, it's not hard to imagine the two men who wrote it having been inspired by marijuana . . .

Epilogue

Food for the body, food for the spirit. After decades of global bans on consuming the plant, the spiritual use of three marijuana preparations connected to the god Shiva continues to be used in India. The weakest is made from leaves (bhang), the one with moderate effects is made from the female flowers (ganja), and the strongest from only the resin that the flowers secrete (charas). The use of these preparations makes it possible to contemplate one's thoughts and sensations, to attain profound meditative states, carry out long spiritual journeys, and reach total dissolution of one's body awareness. It's not surprising, then, that the use of marijuana to help with meditation is connected to the tantric practices of Tibetan Vajrayana Buddhism. As a tool for self-knowledge, marijuana is a powerful teacher of mysteries.

This is what Carl Sagan had to say about marijuana and spirituality:

> I do not consider myself a religious person in
> the usual sense, but there is a religious aspect
> to some highs. The heightened sensitivity in all

areas gives me a feeling of communion with
my surroundings, both animate and inanimate.
Sometimes a kind of existential perception of
the absurd comes over me and I see with awful certainty the hypocrisies and posturing of
myself and my fellow men. And at other times,
there is a different sense of the absurd, a playful and whimsical awareness. Both of these
senses of the absurd can be communicated,
and some of the most rewarding highs I've had
have been in sharing talk and perceptions and
humor. Cannabis brings us an awareness that
we spend a lifetime being trained to overlook
and forget and put out of our minds. A sense of
what the world is really like can be maddening;
cannabis has brought me some feelings for what
it is like to be crazy, and how we use that word
"crazy" to avoid thinking about things that are
too painful for us.[1]

It's worth comparing Sagan's words on the religious effects of marijuana with those of his contemporary Oliver
Sacks: "I was fascinated that one could have such perceptual
changes, and also that they went with a certain feeling of sig-
nificance, an almost numinous feeling. I'm strongly atheist by
disposition, but nonetheless when this happened, I couldn't
help thinking, 'That must be what the hand of God is like.'"[2]

While the use of marijuana for the purpose of inducing
trances, mystical experiences, and religious connections is
very ancient and widespread, there is currently little argument

about the legalization of marijuana for religious use. This is largely due to the stigmatizing of this sacred plant, which to this day has not been granted the proper protections that many countries have given to the ayahuasca vine and leaf, to *Mimosa tenuiflora*, to the peyote cactus, to psilocybin mushrooms, and to many other sacred medicines derived from plants, fungi, and animals. Prohibitionists generally define as drugs those things that other people like and they don't. Shiva's flower, cultivated since Neolithic times with reverence and love, has been sullied in the vilest way to justify the killing and oppression of huge numbers of people. A historic crime that is currently going through a—still erratic— process of reparation. Let us remember the spiritual leader Ras Geraldinho, the elder of a peaceful cannabis church, who, despite his inoffensive gentleness, was imprisoned for almost seven years because of thirty-seven marijuana plants, eventually dying just three years after his release.[3] For having invited Ras Geraldinho to take part in an event on the therapeutic use of marijuana, the scientist Elisaldo Carlini was summoned, at the age of eighty-seven, to give a statement at the sixteenth police precinct of São Paulo.[4] A terrible injustice, against people and the plant.

I've lost count of the days and nights when these flowers have illuminated my path and given me the strength to keep going. These plant teachers have taught me about the many minds in the world, including those that inhabit my own. But to tell that part of the story, we need to go back to the year 2000. By complete chance, I found myself in possession of a book published in 1976 that blew my mind: *The Origin of Consciousness in the Breakdown of the Bicameral Mind* by Julian

Jaynes. Among many other very interesting ideas, the book argues that, in our historic and prehistoric past, altered states of consciousness did not denote illness but rather a special condition that distinguished in a positive way those people who were prone to psychosis and trances and thus highly valued for their divine connections.

For Jaynes, the evolution of the human mind began as a mental space separated into two domains that were distinct but in dialogue. The first domain being the "I" that keeps its attention focused on the present tense, and which, like any mammal, either does or doesn't move at each moment to attain what it desires or avoid what it fears. The second domain, fed by dreams of deceased relatives and friends, is the land of the ancestors, the dwelling of the gods, a collection of voices and images that can influence the first domain with memories of the past and simulations of the future. This mentality, split into two domains, or chambers—hence the term "bicameral"—would have begun hundreds of thousands of years ago, back in the Middle Paleolithic, reaching its peak in the Bronze Age. Around three thousand years ago, however, this mentality that had its basis in hearing divine voices would have collapsed in the face of the difficulties of adapting to the social and environmental complexity caused by the expansion of human culture itself. Today's mentality would be a consequence of that collapse of the two chambers into just one, capable of integrating past, present, and future through an imagination that is free to move in any direction and on any temporal scale. Everybody hears voices, but 99 percent of people believe those voices are their own.

Until I encountered that subversive book, I'd been basically a practicing atheist. As time passed, however, I became ever more curious about the mechanisms that cause various states of consciousness that remained very mysterious, such as dreaming, psychedelia, meditation, hypnosis, mediumistic trance, possession, and psychosis. Jaynes's incredible book made me understand that my brain probably had hardware that was capable of conversing with gods, but my cultural software at the time wasn't allowing me to access that part of the mind. Gods in the brain would be like geraniums in the window box, then: You only need to water them, and they'll blossom.

Challenged by this idea, in mid-2001, I decided to carry out another self-experiment to try to open myself up to other perspectives. Would it be possible to reactivate the gods that were sleeping in the mind of a committed atheist, watering them through propitiatory rituals? In those days, I was living in Durham, North Carolina, with my first partner, neuroscientist Janaina Pantoja. On a trip to São Paulo where I was to give a lecture, I got off the subway at Liberdade to go visit the Santa Rita House of Candles, a lovely store for Afro-Brazilian religious objects. When I left Brazil again, to come back to the U.S., I had in my backpack two beautiful statues of Xangô and Yemanjá. I built an altar, established my Axé, and began to visit those statues daily, intending to communicate with the entities they represented. With a similar intention in mind, I also placed a photo on the altar of my father, who died when I was five years old, as well as other older ancestral symbols, and I began to work. Every day, when I woke up or when I was getting ready for bed, I would stop by there for a moment, trying to make some kind of spiritual connection.

I'll confess I felt a little ridiculous at first, not least because Janaina was doubtful when I showed up with those statues and that idea about the geraniums. Unlike me, she wasn't an atheist at all and had enough experience living with the Umbanda religion to be fearful of playing with the unknown. But I wasn't playing—my attempt was a genuine one, and she ended up accepting what I was trying to do.

I developed my ritual, inventing things using my intuition as a compass and my emotions as a metric. As the days passed, I found myself letting go. I started stretching out my daily altar visit to fifteen minutes or so. And praying quietly, seeking to manifest well-being for those I love and for the planet as a whole. I taught myself how to consecrate pure water, dendê oil (palm oil), incenses, and essential oils, with little theory and a lot of practice. A new combination each day, until I felt that something was stirring within me. A babbling, perhaps the start of an internal conversation?

Yes, though not exactly what I'd been expecting. I started hearing a sarcastic voice that kept repeating, "This is just auto-suggestion, dummy!" At the same time, however, another voice in parallel soothed me: "Don't worry about it—yes, it's probably auto-suggestion but that's fine: What matters is that you keep watering those geraniums."

I felt my capacity for introspection starting to grow and noticed small improvements in motivation, but, frustratingly, the effect was nowhere near a mystical trance. I went on repeating my ritual of visiting the altar, but now I felt ridiculous again. Gradually I became overly conscious of how this would be cause for mockery from most of the people I knew if they could only see me in those moments, with my atheist's cry for the numinous.

I was about to give up on the whole venture when it occurred to me that I might call on the ganja flowers to accompany me on my visits to the altar. An obvious-enough idea, truth be told, but I was so used to encountering ganja only in contexts of recreation and enjoyment that I had somehow neglected this powerful teacher as a possible ally in my spiritual search. I greeted Xangô and Yemanjá, asked permission of my ancestors, and consecrated the ancient sacred weed in a ritual context for the first time . . .

At first, I felt only a profound sense of physical relaxation, accompanied by a delight in existence and an acceleration of my thoughts. I shut my eyes and gradually came to understand that my conscious intention of finding a mystical connection was resonating within me. Not at first directly interacting with any creatures of the mind but in an unashamed liberating of my own life force for a deep dive into the unconscious.

I started to laugh, to sing and dance with no sense of repression at all; I spun around in front of the altar and finally lost all sense of time . . . I felt with every cell in my body that I was having my first-ever experience of a very ancient kind of trance, which our ancestors had experienced countless times ever since the end of the Ice Age. Without really guiding the experience, I caught myself evoking vivid active imaginations throughout the history of my family. And then, without warning, I heard, faintly, a very calm, soothing voice. It was my father—or at least, the father who is alive in me. We hadn't talked in twenty-five years.

"Hello, hello . . ."

"Hi, Dad."

"Hey, son. I'm so glad you called."

"I was missing you."

"And you. How're you doing?"

"Good. I'm happy."

"That's great!"

"What about you?"

"All great—your mind is an absolute carnival, son."

"And how's grandpa?"

"He's losing at cards an awful lot!"

"Hahaha."

"That's all good then. So look, I'll be going now, OK?"

"Stay a bit more."

"I can't now . . . You come with me."

"You go on ahead, I'll be right there."

"Give me a kiss."

smack

I woke from the trance in floods of tears, with the wonderful feeling of being loved and cared for by my dad, who had been lost for so long in the haze of memory . . . Revived by the ganja consecrated at the altar, my relationship was never broken again. With time and practice I also managed to feel the presence of the *orixás* and came to consult them whenever I needed. Today I carry with me a good part of the Yoruba pantheon that has survived in Brazil. Laroiê Exu, Ogunhê Ogum, Odoiá Iemanjá, Atotô Omolu, Kao Kabiecile Xangô, Oke Arô Oxóssi, Epa Baba Oxalá! And many other entities are housed there too—from Aluvaiá of Angola to Ganesha of India, from Asclepius of Greece to Jesus Christ of Galilee. We are a permanent assembly of diverse voices, inherited from our relatives and our cultures. In the garden of a reconnected mind, geraniums thrive, and many other flowers besides.

As long as there is life, we must keep regenerating, re-creating, being reborn. Creating art, science, and solutions for moving forward. An afternoon playing with the kids, a lovely dinner, awesome sex, a great movie, smoking a joint, a beautiful night, passing more easily through each moment, with more gentleness, fewer certainties, more experiences, less automation, more imagination. After all, as Sagan taught us, "The cosmos is within us. We are made of star-stuff. We are a way for the universe to know itself." Any resemblance to the cosmic dance referred to by the philosopher Ailton Krenak is no mere coincidence.[5]

I'm grateful to the flowers for all this and a little more. An increase in perspective, parallax, self-observation of the mind, a seed of conscious navigation. Wandering the paths of the improbable, that which perhaps will never be, and yet maybe, it might, who knows, it just might be . . . and may even grow for all to see. The near impossibility that happened to be a necessity. The outside of any and every closet. The discovery of the space-time discontinuum, as perceived by the brain and its psychological biochemistry. The trigger of what has not been lost. The soul doing mime. I and I.

Acknowledgments

The decision to write this book was made in December 2022, during a Vipassana meditation retreat at the Dhamma Sarana center in Santana de Parnaíba (SP). The text was written in Parnamirim (RN), Rio de Janeiro (RJ), Belo Horizonte (MG), Brasília (DF), and Búzios (RJ). The final version was gathered together and cured between September and October 2023. My thanks are due to so many people who were involved in that process. My apologies if I have forgotten anyone—it might have been the THC . . .

To the intelligence, creativity, rigor, ample repertoire of ideas, clarity, and good humor of the publishers Rita Mattar, Fernanda Diamant, and Eloah Pina, who welcomed the book enthusiastically to Editora Fósforo. To the editorial care of Andressa Veronesi and assistance from Cristiane Alves Avelar. To the copyediting of Gabriela Rocha and Andrea Souzedo, and the layout by Carlos Tranjan of Página Viva. To the detailed research into imagery for the jacket, with Julia Monteiro's art direction and graphic design by Alles Blau. Ani Ganzala Lorde's beautiful artwork on the cover of the Brazilian edition is a delight in itself. I'm so grateful to have had the opportunity to be in conversation with these brilliant gardeners of words and images, which has been such fertile ground for this book to flourish.

To the careful and affectionate devotion of Luiz Schwarcz of Companhia das Letras, who built a bridge to Fósforo.

To the outstanding translation provided by Daniel Hahn, with whom I had the good luck of collaborating once again.

To Júlio Tollendal Gomes Ribeiro, Luiza Mugnol-Ugarte, João Ricardo Lacerda de Menezes, Ester Nakamura-Palácios, Pedro Roitman, Ricardo Teperman, Janaina Pantoja, and Cristiana Schettini Pereira, for their critical reading of different versions of this text.

To Cecília Hedin-Pereira, João Ricardo Lacerda de Menezes, Renato Malcher-Lopes, Cida Carvalho, Fábio Carvalho, Margarete Brito, Marcos Langenbach, Claudio Queiroz, Ricardo Nemer, Emílio Nabas Figueiredo, Renato Filev, Álvaro Monteiro, Simone Leal, Francisco Guimarães, Jorge Quillfeldt, Ester Nakamura-Palácios, Fabrício Pamplona, Tarcísio Barreto Filho, Mano Brown, Marcelo D2, and Jeremy Narbin, for our discussions and for contributing precious testimonies or memories around marijuana.

To Raphael Mechoulam, José Ribeiro-do-Valle, and Elisaldo Carlini, for their intellectual greatness, their moral firmness, their physical courage and geographical boldness in researching marijuana in their own countries, going against the grain of the moral panic of prohibitionism.

To all children and teenagers, who deserve to live and flourish in a world without war, such as Ágatha, Ana, Anny, Bebel, Bela, Bê, Caio, Camila, Cauê, Charlotte, Chico, Clárian, Ernesto, Franziska, Gabi, Gabriel, Isadora, João,

Joca, Juju, Kima, Lara, Leo, Lisa, Lukas, Maria, Mateus, Matias, Pedro, Pietra, Samuca, Sergio, Sofia, Tainá, Thiago, Vico, Xavier.

To the living paragons of Capoeira, the Afro-Indigenous art that defeated prohibition and spread across the planet: Mestres Caxias, João Grande, Paulinho Sabiá, Ramos, Nestor Capoeira, Curumim, Balão, Janja, João Angoleiro, Sabiá da Bahia, Gladson, Roxinho, Jaime de Mar Grande, Marrom, Tati, Marcos, Igor, Alysson, Irani, Ligeirinho, Picapau de Pirangi, Mala Veia e Perninha, as well as Formando Jeguinho, Formanda Lua, Contramestras Flávia Soares, and Contramestre Max.

To the governmental and nongovernmental institutions that promote the legal regulation of marijuana in Brazil, such as the Instituto do Cérebro da Universidade Federal do Rio Grande do Norte, the Grupo de Trabalho sobre Maconha Medicinal da Fiocruz, the Centro de Estudos de Segurança e Cidadania, the Plataforma Brasileira de Política de Drogas, Plataforma Justa, Rede Reforma, the Rede Nacional de Feministas Antiproibicionistas, the Iniciativa Negra por uma Nova Política sobre Drogas, the Rede Pense Livre, the Growroom collective, Fórum Delta9, and the Marcha da Maconha.

To all the patients' associations, so strategically and popularly invaluable, such as Abraflor, Ação Cannabis, Abracam (CE), Abracannabis (RJ), Abrace (PB), Acolher (PE), Acube (SP), Acuca (SP), Ágape (GO), Aliança Verde (DF), Apepi (RJ), Apracam (PR), ArtCanab (GO), Associação Cannabis Medicinal de Rondônia, Bioser (DF), Cannab (BA), Cannape (PE), Cultiva Brasil, Cultive (SP), Curando Ivo

(GO), Curapro (SP), Divina Flor (MS), Federação das Associações de Cannabis Terapêutica (FACT), Flor da Vida (SP), Índica (BA), Liga Cannábica (PB), Mãesconhas do Brasil (SP), Obec (BA), O Saci (SP), Reconstruir (RN), Regenera (AL), Santa Cannabis (SC), Semear (PR), Sociedade Brasileira de Estudos da Cannabis (SP), Sonho Verde Brasil, SouCannabis (GO), Tijucanna (MG), Volta Cannabis (RN).

To all those people who, through their ideas and their actions, have contributed and continue to contribute to the liberation of the flowers, such as Adriana Lamartine, Adriano de Oliveira Carneiro, Adriano Tort, Aílton Krenak, Alcida Menezes, Alexandre de Moraes, Alice Poltosi, Allan Kardec de Barros, Alok, Alvamar Medeiros, Ana Estela Haddad, Ana Hounie, Ana Luiza Greco, Ana Luiza Meira, Ana Priscilla Marinho, Anderson Henrique, Anderson Matos, André Barros, André Ferreira Feiges, André Jung, André Kiepper, Andrea Gallassi, Angelita Araújo, Anielle Franco, Anita Krepp, Anitta, Ann Hedin, Antonio Bittencourt Júnior, Antonio Carlos Moraes, Antonio Nery, Antonio Peticov, Apollo 9, Arnaldo Antunes, Babá Adolfo, Bacalhau, Beatriz Labate, Beatriz Reingenheim, Bela Gil, Bernard Machado, Bezerra da Silva, Bi Ribeiro, Bia Lessa, Black Alien, BNegão, Bob Fernandes, Branco Mello, Brisa, Bruno Gomes, Bruno Lobão, Bruno Santos, Bruno Torturra, Caio Santos Abreu, Camila Leal Ferreira, Canhoto, Carla Coutinho, Carlos José Zimmer Junior, Carolina Nocetti, Casa de Velas Santa Rita, Cássio Yumatã Braz, Cecilia Galício Brandão, Célia Costa Braga, Céu, Charles Gavin, Christiane Tollendal, Christopher Gernand, Cilene Vieira, Cíntia Tollendal, Ciro Pessoa, Clancy Cavnar, Claudia Kober, Cláudia Linhares, Claudine

Ferrão, Claudio Angelo, Clécio Dias, Cleusa Ladário, Clotilde Tânia Rodrigues Luz, Criolo, Cristiano Maronna, Cristiano Simões, Daniel Ganjaman, Daniel Takahashi, Daniela Monteiro, Danilo Thomaz, Dario de Moura, Dartiu Xavier da Silveira, Dayane Guimarães Lima, Débora Sá, Dengue, Denis Petucco, Denis Russo Burgierman, Denise Pires de Carvalho, Dicró, Diego Laplagne, Diogo Busse, Dráulio de Araújo, Drauzio Varella, Dudu Ribeiro, Edi Rock, Edson Fachin, Edson Sarques Prudente, Eduarda Alves Ribeiro, Eduardo Bueno, Eduardo Faveret, Eduardo Sampaio, Eduardo Schenberg, Eduardo Sequerra, Eduardo Suplicy, Edward McRae, Eliana Sousa Silva, Eliane Brum, Eliane Dias, Eliane Nunes, Elisaldo Carlini, Emicida, Emílio Figueiredo, Emílio Vieira, Érico dos Santos Junior, Erivan Melo, Ernesto Saias Soares, Ernesto Soto, Fabio Presgrave, Fabio Toniolo, Fabrício Moreira, Felipe de Castro, Felipe Farias, Felipe Pegado, Fernanda de Almeida, Fernanda Mello, Fernanda Mena, Fernando Antonio Bezerra Tollendal, Fernando Beserra, Fernando Edson Cerqueira Filho, Fernando Gabeira, Fernando Haddad, Fernando Henrique Cardoso, Fernando Moraes, Fernando Velho, MC Fioti, Flávia Ribeiro, Flávio Lobo, Formando Jeguinho (Rafael Bittencourt), Formigão, Francisco de Abreu Franco Netto, Francisco Inácio Bastos, Frederico Prudente, Gabriel Elias, Gabriel Lacombe, Gabriel o Pensador, Gabriela Costa Braga, Gabriela dos Santos, Gabriela Moncau, Gabriela Moraes, Gabriela Oliveira, Gabriela Simão, Gabriella Arima de Carvalho, Geovani Martins, Geraldo Alckmin, Gilberto Dimenstein, Gilberto Gil, Gilmar Bola 8, Gilmar Mendes, Gilson Antunes da Silva, Gira, Glauber Loures, Glauco Tollendal, Glória Maria,

Gregório Duvivier, Guilherme Coelho, Gustavo Da Lua, Hanna Limulja, Hayne Felipe, Helena Borges, Helio Bentes, Hélio Schwartsman, Henrique Carneiro, Henrique Pacheco, Herbert Vianna, Hermano Vianna, Ian Guedes, Ice Blue, Ichiro Takahashi, Igor Praxedes, Ildeu de Castro Moreira, Ilona Szabó, Ingrid Farias, Ingrid Trancoso, Íris Roitman, Isabela Cunha, Isolda Dantas, Ítalo Coelho, Ivan de Araújo, Ivich, Ivo Lopes Araújo, Jackeline Barbosa, Janaína Barboza, Jânio de Freitas, Jefferson Neves Pereira, Jera Guarani Mbya, Jessica Pires, Joana Amador, Joana Prudente, João Barone, João Paulo Costa Braga, João Vieira Jr., Joel Ilan Paciornik, John Fontenele de Araújo, John Lennon, Jorge Du Peixe, José Balestrini, José Daniel Diniz de Melo, José Eduardo Agualusa, José Henrique Torres, José Luís Gomes da Silva, José Paulo, Juliana Borges, Juliana de Araujo Rodrigues, Juliana de Paolinelli, Juliana Lima, Juliana Pimenta, Julio Américo, Julio Delmanto, Karin Moreira, Karla Antunes, Katarina Leão, Katiúscia Ribeiro, Kellen Marques, Kerexu Guarany-Mbya, Kerstin Schmidt, K. L. Jay, Laerte Ladário, Layla Motta, Leandro Pinheiro, Leandro Ramires, Leilane Assunção, Leon Garcia, Leonardo Costa Braga, Leonardo Sinedino, Letícia Simões, Lilia Moritz Schwarcz, Lívia de Melo, Luana Malheiros, Lucas Kastrup, Luciana Boiteux, Luciana de Barros Jaccoud, Luciana Surjus, Luciana Zaffalon, Luciano Arruda, Luciano Ducci, Luciano Roitman, Lúcio Maia, Ludmilla, Luís Eduardo Soares, Luís Fernando Tófoli, Luís Francisco Carvalho, Luís Roberto Barroso, Luísa Tollendal Prudente, Mac Niven, Mãc Beth de Oxum, Mãe Jaci, Mãe Lu, Malu Mader, Mandacaru, Mani de Azevedo, Mapu Huni Kuin, Marcel Grecco, Marcel Segalla Osama,

Marceleza, Marcello Dantas, Marcelo Campos, Marcelo D2, Marcelo Fromer, Marcelo Gomes, Marcelo Grecco, Marcelo Leite, Marcelo Roitman, Marcelo Tas, Marcelo Tollendal Alvarenga, Márcio Dias Gomes, Márcio Roberto, Márcio Sampaio, Marco Marcondes de Moura, Marcos Matias, Marcus Vinicius, Maria Bernardete Cordeiro de Sousa, Maria Carlota Bruno, Maria Carolina Borin, Maria Clara, Maria Lúcia Karam, Maria Rita Kehl, Mariana Alves Ribeiro, Mariana David German, Mariana de Moraes, Mariana Lacerda, Mariana Muniz, Mariano Sigman, Marília Guimarães, Marina Bortolon Moreira, Marina Pádua Reis, Mário Eduardo Pereira, Mário Kertész, Mario Lisboa Theodoro, Mário Moreira, Marisa Mamede, Marisa Monte, Marta Mugnol, Mateus Santana, Matias Maxx, Mauricio Fiore, Maurides Ribeiro, Michele Soares, Millena Machado, Miriam Krenzinger, M. V. Bill, Mychelle Monteiro, Nanda Torres, Nando Reis, Nara Aragão, Natália Bezerra Mota, Nathália Oliveira, Negalê, Nelson Motta, Nice Souza, Nina Kopko, Nísia Trindade, Nobru Pederneiras, Onildo Marini Jr., Orlando Bueno, Orlando Zaccone, Otavio Frias Filho, Padre Ticão, Pamela Pini, Pamella Carvalho, Patrícia Rosa, Patrícia Tollendal Stein, Patrícia Villela Marino, Patrick Coquerel, Paula Dalla'Stella, Paula Signorelli, Paula Zomignani, Paulo Amarante, Paulo de Azevedo, Paulo Fleury, Paulo Gadelha, Paulo Lima, Paulo Mattos, Paulo Miklos, Paulo Teixeira, Paulo Werneck, Pedrinho Moreira e Moabe Filho, Pedro "Pedrada" Caetano, Pedro Andrade, Pedro Bial, Pedro da Costa Mello Neto, Pedro Dória, Pedro Garcia, Pedro Godoy Bueno, Pedro Guinu, Pedro Themóteo Alves Correa, Pedro Zarur, Pepe Mujica, Pertteson Silva,

Priscila Gadelha, Pupillo, Rael, Rafael Franzon, Rafael Kalebe, Ramon Lira, Raphael Ericksen, Raquel Nunes, Ras Geraldinho, Raull Santiago, Rebeca Lerer, Reinaldo Lopes, Reinaldo Takahashi, Renata Monteiro Dantas, Renata Souza, Renato Cinco, Renato Janine Ribeiro, Renato Russo, Ricardo Chaves, Ricardo Ferreira, Ricardo Reis, Richardson Leão, Rita Lee, Roberta Marcondes Costa, Roberta Mugnol de Oliveira, Roberto D'Ávila, Roberto Lent, Rodolfo Variani, Rodrigo Pacheco, Rodrigo Pereira, Rodrigo Quintela, Rodrigo Sampaio, Rolando Monteiro, Ronaldo Bressane, Rosa Weber, Rosane Borges, Rossella Fabri, Samuel Ladário, Sandro Rodrigues, Santos Flores, Sebastián Basalo, Sergio Alves Ribeiro, Sergio Arthuro Motta Rolim, Sérgio Britto, Sergio Guerra, Sergio Neuenschwander, Sergio Ruschi, Sergio Vidal, Seu Jorge, Sheila Geriz, Silvio Almeida, Skunk, Sofia Roitman, Speed, Stevens Rehen, Sueli Carneiro, Sylara Silverio, Tadeu Jungle, Tales Tollendal Alvarenga, Tarciso Velho, Tarsila Tavares, Tarso Araujo, Terezinha Ferreira Mugnol, Tersio Greguol, Thaís Ferreira, Thais Naiara Fonseca, Theo van der Loo, Tiago Albertini Balbino, Tiago Caetano, Toca Ogan, Tom Rocha, Tony Bellotto, Valber Frutuoso, Valcler Rangel, Vera Lúcia Tollendal Gomes Ribeiro, Veronica Nunes, Victor Vilhena Barroso, Vilma Alves Ribeiro, Vincent Brown, Virgínia Carvalho, Viviane Sedola, Wado, Waldemar Magaldi, William Lantelme, Yogi Pinto Pacheco Filho, Yoko Ono, Zé Celso Martinez Correa, Zé Gonzales, and Zeneide Bezerra.

Notes

Marijuana Wins by *Ippon*

1. Francisco Manoel Barroso da Silva, "Descripção de algumas drogas e medicamentos da India, feita em 1799 pelos facultativos de Goa," *Archivo de Pharmacia e Sciencias Accessorias da India Portugueza* vol. 1, no. 12 (1864): pp. 185–91, originally published in 1799, quoted in Chris S. Duvall, "Cannabis and Tobacco in Precolonial and Colonial Africa," *Oxford Research Encyclopedia of African History*, March 29, 2017, doi .org/10.1093/acrefore/9780190277734.013.44.

2. Frank Newport, "Americans and the Future of Cigarettes, Marijuana, Alcohol," Gallup, August 26, 2022, news .gallup.com/opinion/polling-matters/398138/americans -future-cigarettes-marijuana-alcohol.aspx.

3. Cathleen O'Grady, "Cannabis Research Database Shows How U.S. Funding Focuses on Harms of the Drug," *Science*, August 27, 2020, www.science.org/content/article /cannabis-research-database-shows-how-us-funding-focuses -harms-drug.

4. Sanjay Gupta, "Why I Changed My Mind on Weed," CNN Health, August 8, 2013, edition.cnn.com/2013/08 /08/health/gupta-changed-mind-marijuana/index.html.

5. In Washington, the medicinal use of marijuana had been

legalized since 1998. Recreational use was legalized in both states in 2012. "Medical Cannabis. History in Washington," Washington State Department of Health, August 2023, doh.wa.gov/sites/default/files/2023-11/Med Cannabis-HistoryInWashington.pdf.

6. Andrew Daniller, "Two-Thirds of Americans Support Marijuana Legalization," Pew Research Center, November 14, 2019, www.pewresearch.org/short.reads/2019 /11/14/americans-support-marijuana-legalization.

7. Statista, "Cannabis—Worldwide," www.statista.com /outlook/hmo/Cannabis/worldwide; and Fortune Business Insights, "Cannabis Market Size, Share and COVID-19 Impact Analysis," March 17, 2025, www .fortunebusinessinsights.com/industry-reports/cannabis -marijuana-market-100219.

Brazil Is Still Lagging, but It's Making Progress

1. Renato Malcher-Lopes and Sidarta Ribeiro, *Maconha, cérebro e saúde*, preface by João Ricardo Lacerda de Menezes (São Paulo: Reviver, 2019).

2. Marília Juste, "Músico preso por plantar maconha teme que caso se repita," G1, July 17, 2010, g1.globo.com/rio -de-janeiro/noticia/2010/07/musico-preso-por-plantar -maconha-teme-que-caso-se-repita.html.

3. Marília Juste, "Carta sobre descriminalização da maconha divide neurocientistas," G1, July 15, 2010, g1.globo.com /ciencia-e-saude/noticia/2010/07/carta-sobre-descrimi nalizacao-da-maconha-divide-neurocientistas.html.

4. Canal Eu protesto pelo Brasil e Amigos, "Legalização da maconha debate completo Folha de S.Paulo 20/10/2010," September 19, 2012, YouTube video, 58:07, www.youtube .com/watch?v=QOBxnWbS4yI&t=608s.

5. Flávia Cristina, "Estudante comemora autorização para usar remédio à base de maconha," G1, August 29, 2014, g1.globo.com/minas-gerais/noticia/2014/08/estudante-comemora-autorizacao-para-usar-remedio-basede-maconha.html.

6. Camila Brandalise, "Descobri um câncer no parto e a maconha me ajudou a ser mãe de verdade," *Universa*, Uol, November 11, 2018, www.uol.com.br/universa/noticias/redacao/2018/11/08/descobri-um-cancer-no-parto-e-a-maconha-me-ajudou-a-ser-mae-de-verdade.htm.

7. Gabriela Ingrid, "Com esclerose múltipla, me deram só mais cinco anos. Maconha me devolveu a vida," *Viva bem*, Uol, December 7, 2018, www.uol.com.br/vivabem/noticias/redacao/2018/12/07/com-esclerose-multipla-me-deram-cinco-anos-de-vida-maconha-me-salvou.htm.

8. Bruno Levinson, *Baseado em papos reais: maconha* (São Paulo: Blucher, 2023), 162.

9. Cida Carvalho, in discussion with the author, August 1, 2023.

10. "ILEGAL," repense, Mar 27, 2014, YouTube video, 5:40, www.youtube.com/watch?v=CtJJ1pzMKxs.

11. Nelson Marcolin and Ricardo Zorzetto, "Elisaldo Carlini: o uso medicinal da maconha," *Revista Pesquisa*, no. 168 (February 2010), evistapesquisa.fapesp.br/elisaldo-carlini-o-uso-medicinal-da-maconha/.

12. A. W. Zuardi et al., "Action of Cannabidiol on the Anxiety and Other Effects Produced by Delta 9-THC in Normal Subjects," *Psychopharmacology*, Berlin, vol. 76. no. 3 (March 1982): pp. 245–50.

13. Aviva Breuer et al., "Fluorinated Cannabidiol Derivatives: Enhancement of Activity in Mice Models Predictive of Anxiolytic, Antidepressant and Antipsychotic Effects," *PLos One*, vol. 11, no. 7 (July 14, 2016): e0158779.

14. F. S. Guimarães et al., "Antianxiety Effect of Cannabidiol

in the Elevated Plus-Maze," *Psychopharmacology*, Berlin, vol. 100, no. 4 (1990): pp. 558–59.

15. Francisco Guimarães, in discussion with the author, September 2, 2023.

16. E. M. Nakamura et al., "Reversible Effects of Acute and Long-Term Administration of Delta-9-Tetrahydrocannabinol (THC) on Memory in the Rat," *Drug and Alcohol Dependence*, vol. 28, no. 2 (August 1991): pp. 167–75.

17. Ester Nakamura-Palacios, in discussion with the author, August 17, 2023.

18. Fabrício Pamplona, in discussion with the author, August 25, 2023.

19. Jorge Quillfeldt, in discussion with the author, September 4, 2023.

20. Igor Rafael Praxedes de Sales, *Atividade anticrise de fitocomplexos derivados de* Cannabis spp. *E do canabidiol em um modelo de status epilepticus em camundongos* (PhD thesis, doctorate in neuroscience, Natal: UFRN, 2022).

21. Emilio Figueiredo, in discussion with the author, May 14, 2025.

22. Cecília Hedin-Pereira, in discussion with the author, September 18, 2023.

23. "Drauzio Dichava #1/Era uma vez uma planta," Drauzio Varella, April 22, 2019, YouTube video, 7:23, www.youtube.com/watch?v=7fpBrVl883Y.

24. Margarete Brito, in discussion with the author, September 4, 2023.

25. Williane Silva, "Anvisa autoriza cultivo de Cannabis para pesquisa na UFRN," UFRN, December 16, 2022, www.ufrn.br/imprensa/noticias/66596/anvisa-autoriza-cultivo-de-cannabis-para-pesquisa-na-ufrn.

26. Lucas Góis, "Saiba como a UFRN está desenvolvendo um sólido programa de pesquisa sobre Cannabis numa fase pré-clínica," Sechat, February 13, 2023, sechat.com.br

/noticia/saiba-como-a-ufrn-esta-desenvolvendo-um
-solido-programa-de-pesquisa-sobre-cannabis-numa-fase
-pre-clinica.

27. Lester Grinspoon, "Statement of Lester Grinspoon,
MD, Associate Professor of Psychiatry, Harvard Medical
School," Medical Marijuana Referenda Movement in
America: Hearing Before the Subcommittee on Crime
of the Committee on the Judiciary House of Representatives, October 1, 1997, commdocs.house.gov/committees
/judiciary/hju58955.000/hju58955_0f.htm.

The Flower of the Ganges Was Born in China

1. Guanpeng Ren et al., "Large-Scale Whole-Genome Resequencing Unravels the Domestication History of *Cannabis
sativa*," *Science Advances*, vol. 7, no. 29 (July 16, 2021), www
.science.org/doi/10.1126/sciadv.abg2286.

2. Gil Bar-Sela et al., "The Effects of Dosage-Controlled Cannabis Capsules on Cancer-Related Cachexia and Anorexia
Syndrome in Advanced Cancer Patients: Pilot Study," *In-
tegrative Cancer Therapy*, vol. 18 (January–December 2019),
journals.sagepub.com/doi/10.1177/1534735419881498.

3. Robert Ramer et al., "The Endocannabinoid System as a
Pharmacological Target for New Cancer Therapies," *Cancers (Basel)*, vol. 13, no. 22 (November 15, 2021): p. 5701, doi
.org/10.3390/cancers13225701.

4. Anna Maria Malfitano et al., "Update on the Endocannabinoid System as an Anticancer Target," *Expert Opinion
on Therapeutic Targets*, vol. 15, no. 13 (2011): pp. 297–308.
See also Sebastian Sailler et al., "Regulation of Circulating Endocannabinoids Associated with Cancer and Metastases in Mice and Humans," *Oncoscience*, vol. 1, no. 4
(April 30, 2014): pp. 272–82.

5. Robert Ramer and Burkhard Hinz, "Chapter Twelve: Cannabinoids as Anticancer Drugs," *Advances in Pharmacology*, vol. 80 (2017): pp. 397–436. See also Guillermo Velasco et al., "Towards the Use of Cannabinoids as Antitumour Agents," *Nature Reviews Cancer*, vol. 12 (May 4, 2012): pp. 436–44; and Sean D. McAllister et al., "Cannabinoid Cancer Biology and Prevention," *Journal National Cancer Institute Monographs*, no. 58 (December 2010): pp. 99–106.

6. Hadassah Medical Organization, "A Study: Pure CBD as Single-Agent for Solid Tumor," Clinical Trials, September 2014, classic.clinicaltrials.gov/ct2/show/nct02255292; and Jazz Pharmaceuticals, "A Safety Study of Sativex in Combination with Dose-Intense Temozolomide in Patients with Recurrent Glioblastoma," Clinical Trials, December 2022, classic.clinicaltrials.gov/ct2/show/nct01812603.

7. Julian Kenyon et al., "Report of Objective Clinical Responses of Cancer Patients to Pharmaceutical-Grade Synthetic Cannabidiol," *Anticancer Research*, vol. 38, no. 10 (October 2018): pp. 5831–35.

8. Hui-Lin Li, "An Archaeological and Historical Account of Cannabis in China," *Economic Botany*, vol. 28, no. 4 (October–December 1974): pp. 437–48, www.jstor.org/stable/4253540.

9. Mia Touw, "The Religious and Medicinal Uses of *Cannabis* in China, India and Tibet," *Journal of Psychoactive Drugs*, vol. 13, no. 1 (January–March 1981): pp. 23–34. See also Antonio Waldo Zuardi, "History of Cannabis as a Medicine: A Review," *Brazilian Journal of Psychiatry*, vol. 28, no. 2 (June 2006): pp. 153–57.

10. Fan Ka Wai, "On Hua Tuo's Position in the History of Chinese Medicine," *The American Journal of Chinese Medicine*, vol. 32, no. 2 (2004): pp. 313–20.

11. The general sacramental use of marijuana in the

Rastafarian religion reflects the cultural contact between Afro-descendants and Indian immigrants in Jamaica in the nineteenth and twentieth centuries. See Vincent E. Burgess, "Indian Influences on Rastafarianism" (thesis, Ohio State University, 2007), kb.osu.edu/server/api/core /bitstreams/b0a91dae-f4a3-57e4-9d66-32c291dad7e8 /content.

12. Christian Rätsch, *Plants of Love* (Berkeley: Ten Speed Press, 1997), 82, 86.

13. Dominik Wujastyk, "*Cannabis* in Traditional Indian Herbal Medicine," in *Ayurveda at the Crossroads of Care and Cure: Proceedings of the Indo-European Seminar on Ayurveda*, ed. A. Salema (Lisboa: Universidade Nova de Lisboa, 2002), 45–73.

14. Robert C. Clarke and Mark D. Merlin, *Cannabis: Evolution and Ethnobotany* (Berkeley: University of California Press, 2016), 234.

15. F. Parsche et al., "Drugs in Ancient Populations," *Lancet*, vol. 341, no. 8843 (1993): p. 503.

16. Brian M. du Toit, "Man and Cannabis in Africa: A Study of Diffusion," *African Economic History*, no. 1 (Spring 1976): pp. 17–35.

17. Zach Fenech, "73 Weed Quotes from Influential Voices Throughout History," *Herb*, July 25, 2024, herb.co/news /culture/weed-quotes/.

18. "Linho Cânhamo," *O Arquivo Nacional e a História Luso-Brasileira*, November 29, 2021, historialuso.arquivonacional .gov.br/index.php?option=com_content&view=article&id =6444:linho-canhamo&catid=2080&Itemid=215.

19. Edward MacRae and Wagner Coutinho Alves, eds., *Fumo de Angola: Cannabis, racismo, resistência cultural e espiritualidade* (Salvador: Edufba, 2016). See also Luísa Saad, *Fumo de negro: a criminalização da maconha no pós-abolição* (Salvador: Edufba, 2019); and Chris S. Duvall, *The African Roots of Marijuana* (Durham, NC: Duke University Press, 2019).

20. Laurentino Gomes, *Escravidão: do primeiro leilão de cativos em Portugal até a morte de Zumbi dos Palmares*, vol. 1 (Rio de Janeiro: Globo, 2019).

21. F. de Assis Iglésias, "Sôbre o vício da diamba," in *Brasil, Comissão Nacional de Fiscalização de Entorpecentes. Maconha: coletânea de trabalhos brasileiros*, 2nd ed. (Rio de Janeiro: Serviço Nacional de Educação Sanitária, 1958), 15–23.

22. Leonard E. Barrett, *The Rastafarians* (Boston: Beacon Press, 1997); and Ennis B. Edmonds, *Rastafari: A Very Short Introduction* (Oxford: Oxford University Press, 2012).

23. On indigenous habits relating to marijuana, see Eduardo Galvão and Charles Wagley, *Os índios Tenetehara: uma cultura em transição* (Rio de Janeiro/Distrito Federal: Dep. Imprensa Nacional/Ministério da Educação e Cultura, 1961); Anthony R. Henman, "A guerra às drogas é uma guerra etnocida: um estudo do uso da maconha entre os indígenas tenetehara do Maranhão," *Religião e Sociedade*, Rio de Janeiro, vol. 10 (November 1983): pp. 37–48; Guilherme Pinho, "Medicinas da floresta: conexões e conflitos cosmo-ontológicos," *Horizontes Antropológicos*, vol. 51 (2018): pp. 229–58; and Marcos Pivetta, "As lições dos Kraho," *Revista Pesquisa*, São Paulo, Fapesp, no. 70 (2001), revistapesquisa.fapesp.br/as-licoes-dos-kraho/.

24. Guanpeng Ren et al., "Large-Scale Whole-Genome Resequencing Unravels the Domestication History of *Cannabis sativa*," *Science Advances* 7, no. 29 (July 16, 2021), www.science.org/doi/10.1126/sciadv.abg2286.

25. Ailton Krenak, *Ideias para adiar o fim do mundo* (São Paulo: Companhia das Letras, 2020).

The Science of the Flowers

1. Sir William Brooke O'Shaughnessy, "On the Preparation of the Indian Hemp, or Guijah," *Journal of the Asiatic*

Society of Bengal, no. 93 (September 1839): pp. 732–45, archive.org/details/journalofasiatic08asia/page/732 /mode/1up?view=theater.

2. Charles Sajou, *Sajou's Analytic Cyclopedia of Practical Medicine. Cannabis Indica to Dermatitis*, vol. 3 [1922] (London: Forgotten Books, 2018).

3. Lucas V. Araújo Silva, *Catalogo de extractos fluidos* (Rio de Janeiro: Silva Araujo & Cia., 1930).

4. Elisaldo Araújo Carlini, "A história da maconha no Brasil," *Jornal Brasileiro de Psiquiatria*, vol. 55, no. 4 (2006), doi .org/10.1590/S0047-20852006000400008.

5. Y. Gaoni and R. Mechoulam, "Isolation, Structure, and Partial Synthesis of an Active Constituent of Hashish," *Journal of the American Chemical Society*, vol. 86, no. 8 (April 1, 1964): pp. 1646–47.

6. J. M. Cunha et al., "Chronic Administration of Cannabidiol to Healthy Volunteers and Epileptic Patients," *Pharmacology*, vol. 21, no. 3 (1980): pp. 175–85.

7. The studies that describe these trials are: O. Devinsky et al., "Trial of Cannabidiol for Drug-Resistant Seizures in the Dravet Syndrome. Cannabidiol in Dravet Syndrome Study Group," *The New England Journal of Medicine*, vol. 376, no. 21 (May 25, 2017): pp. 2011–20; and O. Devinsky et al., "Effect of Cannabidiol on Drop Seizures in the Lennox-Gastaut Syndrome. GWPCARE3 Study Group," *The New England Journal of Medicine*, vol. 378, no. 20 (May 16, 2018): pp. 1888–97.

8. W. A. Devane et al., "Isolation and Structure of a Brain Constituent That Binds to the Cannabinoid Receptor," *Science*, vol. 258, no. 5090 (December 1992): pp. 1946–49.

9. David Robbe et al., "Cannabinoids Reveal Importance of Spike Timing Coordination in Hippocampal Function," *Nature Neuroscience*, vol. 9, no. 12 (November 19, 2006): pp. 1526–33.

10. Paulo Fleury-Teixeira et al., "Effects of CBD-Enriched

Cannabis sativa Extract on Autism Spectrum Disorder Symptoms: An Observational Study of 18 Participants Undergoing Compassionate Use," *Frontiers in Neurology*, vol. 10 (October 30, 2019): p. 1145.

11. Sara Lukmanji et al., "The Co-Occurrence of Epilepsy and Autism: A Systematic Review," *Epilepsy & Behavior*, vol. 98 (September 2019): pp. 238–48.

Yang, Yin, and Many Other Molecules Besides

1. W. A. Devane et al., "Determination and Characterization of a Cannabinoid Receptor in Rat Brain," *Molecular Pharmacology*, vol. 34, no. 5 (November 1988): pp. 605–13. See also Lisa A. Matsuda et al., "Structure of a Cannabinoid Receptor and Functional Expression of the Cloned cDNA," *Nature*, vol. 346 (August 9, 1990): pp. 561–64.

2. J. M. Derocq et al., "Cannabinoids Enhance Human B-Cell Growth at Low Nanomolar Concentrations," *FEBS Letters*, vol. 369, no. 2–3 (August 7, 1995): pp. 177–82. See also M. Bouaboula et al., "Cannabinoid-Receptor Expression in Human Leukocytes," *European Journal of Biochemistry*, vol. 214, no. 1 (May 1993): pp. 173–80.

3. Ethan B. Russo et al., "Survey of Patients Employing Cannabigerol-Predominant Cannabis Preparations: Perceived Medical Effects, Adverse Events, and Withdrawal Symptoms," *Cannabis and Cannabinoid Research*, vol. 7, no. 5 (October 12, 2022): pp. 706–16.

4. Daniel I. Brierley et al., "Cannabigerol Is a Novel, Well-Tolerated Appetite Stimulant in Pre-Satiated Rats," *Psychopharmacology*, vol. 233 (2016): pp. 3603–13. See also Daniel I. Brierley et al., "A Cannabigerol-Rich *Cannabis sativa* Extract, Devoid of Δ9-Tetrahydrocannabinol, Elicits

Hyperphagia in Rats," *Behavioural Pharmacology*, vol. 28, no. 4 (June 2017): pp. 280–84.

5. Daniel I. Brierley et al., "Chemotherapy-Induced Cachexia Dysregulates Hypothalamic and Systemic Lipoamines and Is Attenuated by Cannabigerol," *Journal of Cachexia, Sarcopenia and Muscle*, vol. 10, no. 4 (August 2019): pp. 844–59.

6. Michael R. Irwin and Michael V. Vitiello, "Implications of Sleep Disturbance and Inflammation for Alzheimer's Disease Dementia," *The Lancet Neurology*, vol. 18, no. 3 (March 2019): pp. 296–306; Ehsan Shokri-Kojori et al., "Beta-Amyloid Accumulation in the Human Brain After One Night of Sleep Deprivation," *PNAS*, vol. 115, no. 17 (April 9, 2018): pp. 4483–88; Edwin E. Martínez Leo and Maira R. Segura Campos, "Effect of Ultra-Processed Diet on Gut Microbiota and Thus Its Role in Neurodegenerative Diseases," *Nutrition*, vol. 71 (March 2020): p. 110609; Zurine De Miguel et al., "Exercise Plasma Boosts Memory and Dampens Brain Inflammation via Clusterin," *Nature*, vol. 600 (2021): pp. 494–99; and Longfei Xu et al., "Treadmill Exercise Promotes E3 Ubiquitin Ligase to Remove Amyloid Beta and P-tau and Improve Cognitive Ability in APP/PS1 Transgenic Mice," *Journal of Neuroinflammation*, vol. 19 (2022): p. 243.

7. Huiping Li et al., "Association of Ultraprocessed Food Consumption with Risk of Dementia: A Prospective Cohort Study," *Neurology*, vol. 99, no. 10 (September 6, 2022): pp. e1056–66.

8. Antonio Currais et al., "Amyloid Proteotoxicity Initiates an Inflammatory Response Blocked by Cannabinoids," *npj Aging and Mechanisms of Disease*, vol. 2 (2016): p. 16012; and Zhibin Liang et al., "Cannabinol Inhibits Oxytosis/Ferroptosis by Directly Targeting Mitochondria Independently

of Cannabinoid Receptors," *Free Radical Biology Medicine*, vol. 180 (February 20, 2022): pp. 33–51.

9.   Sophie Watts et al., "Cannabis Labelling Is Associated with Genetic Variation in Terpene Synthase Genes," *Nature Plants*, vol. 7 (2021): pp. 1330–34.

10. Ethan B. Russo, "Taming THC: Potential Cannabis Synergy and Phytocannabinoid-Terpenoid Entourage Effects," *British Journal of Pharmacology*, vol. 163, no. 7 (August 2011): pp. 1344–64. See also Rita De Cássia da Silveira e Sá et al., "Analgesic-Like Activity of Essential Oil Constituents: An Update," *International Journal of Molecular Sciences*, vol. 18, no. 12 (2017): p. 2392; and Hannah M. Harris et al., "Role of Cannabinoids and Terpenes in Cannabis-Mediated Analgesia in Rats," *Cannabis and Cannabinoid Research*, vol. 4, no. 3 (2019): pp. 177–82.

11. On the reduction of cytokines, see Patrizia A. D'Alessio et al., "Oral Administration of *d*-Limonene Controls Inflammation in Rat Colitis and Displays Anti-Inflammatory Properties as Diet Supplementation in Humans," *Life Sciences*, vol. 92, no. 24–26 (July 10, 2013): pp. 1151–56. On scarring effects, see Patrizia A. D'Alessio et al., "Skin Repair Properties of *d*-Limonene and Perillyl Alcohol in Murine Models," *Anti-Inflammatory and Anti-Allergy Agents in Medicinal Chemistry*, vol. 13, no. 1 (2014): pp. 29–35. On antidepressant effects, see Zahra Lorigooini et al., "Limonene Through Attenuation of Neuroinflammation and Nitrite Level Exerts Antidepressant-Like Effect on Mouse Model of Maternal Separation Stress," *Behavioural Neurology* (January 29, 2021): p. 8817309.

12. Juyong Kim et al., "The Cannabinoids, CBDA and THCA, Rescue Memory Deficits and Reduce Amyloid-Beta and Tau Pathology in an Alzheimer's Disease-Like Mouse Model," *International Journal of Molecular Sciences*, vol. 24, no. 7 (April 6, 2023): p. 6827.

13. Shimon Ben-Shabat et al., "An Entourage Effect: Inactive Endogenous Fatty Acid Glycerol Esters Enhance 2-Arachidonoyl-Glycerol Cannabinoid Activity," *European Journal of Pharmacology*, vol. 353, no. 1 (July 17, 1998): pp. 23–31.

14. Sari G. Ferber et al., "The 'Entourage Effect': Terpenes Coupled with Cannabinoids for the Treatment of Mood Disorders and Anxiety Disorders," *Current Neuropharmacology*, vol. 18, no. 2 (2020): pp. 87–96.

15. Henry Blanton et al., "Cannabidiol and Beta-Caryophyllene in Combination: A Therapeutic Functional Interaction," *International Journal of Molecular Sciences*, vol. 23, no. 24 (2022): p. 15470.

16. Noa Raz et al., "Terpene-Enriched CBD Oil for Treating Autism-Derived Symptoms Unresponsive to Pure CBD: Case Report," *Frontiers in Pharmacology*, vol. 13 (2022): p. 979403.

17. Shimrit Uliel-Sibony et al., "Cannabidiol-Enriched Oil in Children and Adults with Treatment-Resistant Epilepsy: Does Tolerance Exist?," *Brain & Development*, vol. 43, no. 1 (January 2021): pp. 89–96.

18. Ailim Cabral, "A história de vida . . . ," *Correio Braziliense*, November 21, 2021, www.correiobraziliense.com.br /revista-do-correio/2021/11/4963751-vida-e-ciencia.html.

19. Patrícia Montagner et al., "Individually Tailored Dosage Regimen of Full-Spectrum Cannabis Extracts for Autistic Core and Comorbid Symptoms: A Real-Life Report of Multi-Symptomatic Benefits," *Frontiers in Psychiatry*, vol. 14 (August 20, 2023).

20. Leandro Ramires, "Uso medicinal da cannabis no país," *Estado de Minas*, June 13, 2023, www.em.com.br/app /noticia/opiniao/2023/06/12/interna_opiniao,1505745 /uso-medicinal-da-cannabis-no-pais.shtml?fbclid=PAAa bEC6t99T1BYkkSDVlbAHUxZVROIlQFhlhj3Yk0I Qtf_B8lC3gN01e1Xp8.

21. Lihi Bar-Lev Schleider et al., "Real Life Experience of Medical Cannabis Treatment in Autism: Analysis of Safety and Efficacy," *Scientific Reports*, vol. 9 (January 17, 2019): p. 200. See also Dana Barchel et al., "Oral Cannabidiol Use in Children with Autism Spectrum Disorder to Treat Related Symptoms and Co-Morbidities," *Frontiers*, vol. 9 (January 8, 2019): p. 1521; and Micha Hacohen et al., "Children and Adolescents with ASD Treated with CBD-Rich Cannabis Exhibit Significant Improvements Particularly in Social Symptoms: An Open Label Study," *Translational Psychiatry*, vol. 12 (2022): p. 375.

22. Bláthnaid McCoy et al., "A Prospective Open-Label Trial of a CBD/THC Cannabis Oil in Dravet Syndrome," *Annals of Clinical and Translational Neurology*, vol. 5, no. 9 (September 2018): pp. 1077–88; and Lizzie Wade, "Legal Highs Make Uruguay a Beacon for Marijuana Research," *Science*, vol. 344, no. 6189 (June 13, 2014): p. 1217.

23. Luigia Cristino et al., "Cannabinoids and the Expanded Endocannabinoid System in Neurological Disorders," *Nature Reviews Neurology*, vol. 16 (2020): pp. 9–29.

Marijuana Doesn't Kill Neurons, It Makes Them Flourish

1. "Campanha contra uso de maconha retrata usuário como bicho-preguiça," Uol, December 26, 2015, noticias.uol .com.br/internacional/ultimas-noticias/2015/12/26 /campanha-contra-uso-de-maconha-retrata-usuario-como -bicho-preguica.htm.

2. Júlia Portela, "Polícia apreende nota de R$420 com bicho-preguiça e folha de maconha," *Universo Online*, November 25, 2021, noticias.uol.com.br/cotidiano/ultimas-noticias/2024 /07/04/nota-420-maconha-bicho-preguica-pr.htm.

3. "Deputado Laerte Bessa indenizará governador Rollemberg por ofendê-lo-em discursos," Migalhas, April 26, 2017, www.migalhas.com.br/quentes/257864 /deputado-laerte-bessa-indenizara-governador-rollem bergpor-ofende-lo-em-discursos.

4. Mohini Ranganathan and Deepak Cyril D'Souza, "The Acute Effects of Cannabinoids on Memory in Humans: A Review," *Psychopharmacology*, vol. 188 (2006): pp. 425–44. See also Kirsten C. S. Adam et al., "Delta-9-Tetrahydrocannabinol (THC) Impairs Visual Working Memory Performance: A Randomized Crossover Trial," *Neuropsychopharmacology*, vol. 45, no. 11 (October 2020): pp. 1807–16.

5. Rachel Lees et al., "Effect of Four-Week Cannabidiol Treatment on Cognitive Function: Secondary Outcomes from a Randomised Clinical Trial for the Treatment of Cannabis Use Disorder," *Psychopharmacology*, vol. 240, no. 2 (February 2023): pp. 337–46.

6. Celia J. A. Morgan et al., "Impact of Cannabidiol on the Acute Memory and Psychotomimetic Effects of Smoked Cannabis: Naturalistic Study," *British Journal of Psychiatry*, vol. 197, no. 4 (October 2010): pp. 285–90.

7. Eduardo Vanini, "'Tenho uma memória incrível, não sei por quê. Fumo maconha todos os dias, há 55 anos', diz Nelson Motta," *O Globo*, October 12, 2019, oglobo.globo .com/ela/gente/tenho-uma-memoria-incrivel-nao-sei -porque-fumo-maconha-todos-os-dias-ha-55-anos-diz -nelson-motta-24013953.

8. Guang Yang et al., "Sleep Promotes Branch-Specific Formation of Dendritic Spines After Learning," *Science*, vol. 344, no. 6188 (June 6, 2014): pp. 1173–78. See also Wei Li et al., "REM Sleep Selectively Prunes and Maintains New Synapses in Development and Learning," *Nature Neuroscience*, vol. 20 (2017): pp. 427–37.

9.   Kirsty L. Spalding et al., "Dynamics of Hippocampal Neurogenesis in Adult Humans," *Cell*, vol. 153, no. 6 (June 6, 2013): pp. 1219–27.

10. Shawn F. Sorrells et al., "Human Hippocampal Neurogenesis Drops Sharply in Children to Undetectable Levels in Adults," *Nature*, vol. 555, no. 7696 (2018): pp. 377–81.

11. Kunlin Jin et al., "Defective Adult Neurogenesis in CB1 Cannabinoid Receptor Knockout Mice," *Molecular Pharmacology*, vol. 66, no. 2 (August 2004): pp. 204–8.

12. Wen Jiang et al., "Cannabinoids Promote Embryonic and Adult Hippocampus Neurogenesis and Produce Anxiolytic- and Antidepressant-Like Effects," *The Journal of Clinical Investigation*, vol. 115, no. 11 (November 2005): pp. 3104–16.

13. Laura Micheli et al., "Depression and Adult Neurogenesis: Positive Effects of the Antidepressant Fluoxetine and of Physical Exercise," *Brain Research Bulletin*, vol. 143 (October 2018): pp. 181–93.

14. Andras Bilkei-Gorzo et al., "A Chronic Low Dose of Delta-9-Tetrahydrocannabinol (THC) Restores Cognitive Function in Old Mice," *Nature Medicine*, vol. 23, no. 6 (2017): pp. 782–87.

15. Joanna A. Komorowska-Müller et al., "Chronic Low-Dose Delta-9-Tetrahydrocannabinol (THC) Treatment Stabilizes Dendritic Spines in 18-Month-Old Mice," *Scientific Reports*, vol. 13, no. 1 (2023): p. 1390.

16. Andreas Zimmer et al., "Increased Mortality, Hypoactivity, and Hypoalgesia in Cannabinoid CB1 Receptor Knockout Mice," *PNAS*, vol. 96, no. 10 (May 11, 1999): pp. 5780–85.

17. F. Berrendero et al., "Changes in Cannabinoid Receptor Binding and mRNA Levels in Several Brain Regions of Aged Rats," *Biochimica et Biophysica Acta*, vol. 1407, no. 3 (September 30, 1998): pp. 205–14.

18. Anastasia Piyanova et al., "Age-Related Changes in the Endocannabinoid System in the Mouse Hippocampus," *Mechanisms of Ageing and Development*, vol. 150 (September 2015): pp. 55–64.

19. Prakash Nidadavolu et al., "Dynamic Changes in the Endocannabinoid System during the Aging Process: Focus on the Middle-Age Crisis," *International Journal of Molecular Sciences*, vol. 23, no. 18 (September 6, 2022): p. 10254.

20. Prakash Nidadavolu et al., "Efficacy of Delta-9-Tetrahydrocannabinol (THC) Alone or in Combination with a 1:1 Ratio of Cannabidiol (CBD) in Reversing the Spatial Learning Deficits in Old Mice," *Frontiers in Aging Neuroscience*, vol. 13 (August 29, 2021): p. 718850.

21. Carl Sagan and Lester Grinspoon, *Marihuana Reconsidered* (Cambridge: Harvard University Press, 1971).

Living with the Flowers

1. David Graeber and David Wengrow, *The Dawn of Everything: A New History of Humanity* (London: Allen Lane, 2021).

2. Tor D. Wager and Lauren Y. Atlas, "The Neuroscience of Placebo Effects: Connecting Context, Learning and Health," *Nature Reviews Neuroscience*, vol. 16, no. 7 (2015): pp. 403–18.

3. Wager and Atlas, "The Neuroscience of Placebo Effects."

4. Dan Jin et al., "Secondary Metabolites Profiled in Cannabis Inflorescences, Leaves, Stem Barks, and Roots for Medicinal Purposes," *Scientific Reports*, vol. 10 (February 24, 2020): p. 3309.

5. Dante F. Placido and Charles C. Lee, "Potential of Industrial Hemp for Phytoremediation of Heavy Metals," *Plants (Basel)*, vol. 11, no. 5 (February 23, 2022): p. 595. See also

Evangelia E. Golia et al., "Investigating the Potential of Heavy Metal Accumulation from Hemp. The Use of Industrial Hemp (*Cannabis Sativa L.*) for Phytoremediation of Heavily and Moderated Polluted Soils," *Sustainable Chemistry and Pharmacy*, vol. 31 (April 2023): p. 100961; and Yudi Wu et al., "Phytoremediation of Contaminants of Emerging Concern from Soil with Industrial Hemp (*Cannabis sativa L.*): A Review," *Environment Development and Sustainability*, vol. 23, no. 2 (October 2021).

6.   Masashi Soga et al., "Gardening Is Beneficial for Health: A Meta-Analysis," *Preventive Medicine Reports*, vol. 5 (March 2017): pp. 92–99. See also Michelle Howarth et al., "What Is the Evidence for the Impact of Gardens and Gardening on Health and Well-Being: A Scoping Review and Evidence-Based Logic Model to Guide Healthcare Strategy Decision Making on the Use of Gardening Approaches as a Social Prescription," *BMJ Open*, vol. 10 (July 19, 2020): e036923.

7.   Oliver Sacks, "Why We Need Gardens," In *Everything in Its Place: First Loves and Last Tales* (New York: Knopf, 2019).

8.   Carl A. Roberts et al., "Exploring the Munchies: An Online Survey of Users' Experiences of Cannabis Effects on Appetite and the Development of a Cannabinoid Eating Experience Questionnaire," *Journal of Psychopharmacology*, vol. 33, no. 9 (September 2019): pp. 1149–59.

9.   Marco Koch et al., "Hypothalamic POMC Neurons Promote Cannabinoid-Induced Feeding," *Nature*, vol. 519, no. 7541 (February 18, 2015): pp. 45–50. See also Shi Di et al., "Nongenomic Glucocorticoid Inhibition via Endocannabinoid Release in the Hypothalamus: A Fast Feedback Mechanism," *Journal of Neuroscience*, vol. 23, no. 12 (June 15, 2003): pp. 4850–57; and Renato Malcher-Lopes et al., "Opposing Crosstalk Between Leptin and Glucocorticoids Rapidly Modulates Synaptic Excitation via

Endocannabinoid Release," *Journal of Neuroscience*, vol. 26, no. 24 (June 14, 2006): pp. 6643–50.

10. Carl Sagan and Lester Grinspoon, *Marihuana Reconsidered* (Cambridge: Harvard University Press, 1971).

11. M. A. De Luca et al., "Cannabinoid Facilitation of Behavioral and Biochemical Hedonic Taste Responses," *Neuropharmacology*, vol. 63, no. 1 (July 2012): pp. 161–68.

12. Edgar Soria-Gómez et al., "The Endocannabinoid System Controls Food Intake via Olfactory Processes," *Nature Neurosciences*, vol. 17, no. 3 (2014): pp. 407–15.

13. "Cannabis Cuisine," CBS Sunday Morning, November 20, 2022, YouTube video, 4:10, www.youtube.com /watch?v=nYLOeczyDHU&t=34s.

14. Whitney L. Ogle et al., "How and Why Adults Use Cannabis During Physical Activity," *Journal of Cannabis Research*, vol. 4 (May 18, 2022): p. 24.

15. "Inflamação e a dor de atletas tratados com Cannabis," Cannabis & Saúde, July 21, 2021, YouTube video, 1:08:21, www.youtube.com/watch?v=V6_YbspSdkU&t=627s.

16. Jeff Tracy, "Where It Stands: Weed Policies by U.S. Sports League," Axios, October 20, 2021, www.axios .com/2021/10/20/weed-policies-sports-leagues-nba-mlb -nfl-nhl.

17. Louise Gwilliam, "Cannabis and Sport: NBA Winner Matt Barnes 'Smoked Before Games,'" BBC Sport, May 31, 2018, www.bbc.com/sport/basketball/43836214.

18. Leah Gillett et al., "Arrhythmic Effects of Cannabis in Ischemic Heart Disease," *Cannabis and Cannabinoid Research*, vol. 8, no. 5 (October 9, 2023): pp. 867–76.

19. Ellen Wiebe and Alanna Just, "How Cannabis Alters Sexual Experience: A Survey of Men and Women," *Journal of Sexual Medicine*, vol. 16, no. 11 (November 2019): pp. 1758–62.

20. Andrea Donatti Gallassi et al., "The Increased Alcohol and Marijuana Use Associated with the Quality of Life

and Psychosocial Aspects: A Study During the COVID-19 Pandemic in a Brazilian University Community," *International Journal of Mental Health and Addiction*, vol. 22 (October 21, 2022): pp. 1–21.

21. Amanda Moser et al., "The Influence of Cannabis on Sexual Functioning and Satisfaction," *Journal of Cannabis Research*, vol. 5 (January 20, 2023): p. 2.

22. Sagan and Grinspoon, *Marihuana Reconsidered*.

23. Hui-Lin Li, "The Origin and Use of Cannabis in Eastern Asia Linguistic-Cultural Implications," *Economic Botany*, vol. 28, no. 3 (July–September 1974): pp. 293–301; Kazuo Yoshihara and Yoshio Hirose, "The Sesquiterpenes of Ginseng," *Bulletin of the Chemical Society of Japan*, vol. 48, no. 7 (received 1975; published 2006): pp. 2078–80. See also Rita Richter et al., "Three Sesquiterpene Hydrocarbons from the Roots of *Panax ginseng* C.A. Meyer (Araliaceae)," *Phytochemistry*, vol. 66, no. 23 (December 2005): pp. 2708–13; and Hui-Lin Li, "An Archaeological and Historical Account of Cannabis in China," *Economic Botany*, vol. 28, no. 4 (October–December 1974): pp. 437–48, www.jstor.org/stable/4253540.

24. Zerrin Atakan et al., "The Effect of Cannabis on Perception of Time: A Critical Review," *Current Pharmaceutical Design*, vol. 18, no. 32 (2012): pp. 4915–22.

25. Anna Muro et al., "Cannabis and Its Different Strains," *Experimental Psychology*, vol. 68, no. 2 (March 2021): pp. 57–66.

26. R. Andrew Sewell et al., "Acute Effects of THC on Time Perception in Frequent and Infrequent Cannabis Users," *Psychopharmacology*, vol. 226 (2013): pp. 401–13.

27. Ana Paula Francisco et al., "Cannabis Use in Attention-Deficit/Hyperactivity Disorder (ADHD): A Scoping Review," *Journal of Psychiatric Research*, vol. 157 (January 2023): pp. 239–56.

28. Fernando de Barros e Silva, "A última entrevista de Otavio Frias Filho," *Folha de S.Paulo*, September 23, 2018, www1

.folha.uol.com.br/ilustrissima/2018/09/a-ultima-entrevista
-de-otavio-frias-filho.shtml.

29. Bruno Levinson, *Baseado em papos reais: maconha* (São Paulo: Blucher, 2023), 20–21.

Free Canine-nabis!

1. Callie Barrons, "25 Inspirational Quotes About Weed," *High Times*, archived January 22, 2021, at web.archive.org /web/20210122151320/https://hightimes.com/culture /inspirational-quotes-about-weed/.
2. Barrons, "25 Inspirational Quotes About Weed."
3. Ellen Komp, "Cannabis and 'Muggles': An Etymology," Leafly, November 13, 2018, www.leafly.com/news /lifestyle/etymology-of-muggle-mawrijuana.
4. High Times, "High Times Greats: Louis Armstrong," *High Times*, archived December 17, 2024, at web.archive .org/web/20241217173410/https://hightimes.com /culture/high-times-greats-louis-armstrong/.
5. Zach Fenech, "73 Weed Quotes from Influential Voices Throughout History," *Herb*, July 25, 2024, herb.co/news /culture/weed-quotes/. See also Barrons, "25 Inspirational Quotes About Weed."
6. Bruno Levinson, *Baseado em papos reais: maconha* (São Paulo: Blucher, 2023), 129.
7. Mano Brown, "Entrevista com Sidarta Ribeiro," *Mano a mano*, season 2, episode 5, April 21, 2022, open.spotify .com/episode/0ml5hCcOCu3LP5Pt2wkFnZ?si=43dabf 3382f8440f.
8. Phie Jacobs, "Researchers Applaud Health Officials' Push to Ease Cannabis Restrictions," *Science*, September 1, 2023, www.science.org/content/article/researchers -applaud-hhs-push-ease-marijuana-restrictions.

9.   Lois Beckett, "San Francisco Backs Reparations Plans, Including $5M to Eligible Black Adults," *The Guardian*, March 14, 2023, www.theguardian.com/us-news/2023 /mar/14/san-francisco-reparation-plans-black-residents.

10. Francisco Inácio Pinkusfeld Monteiro Bastos et al., "III Levantamento Nacional sobre o Uso de Drogas pela População Brasileira," *Repositório Institucional da Fiocruz*, Instituto de Comunicação e Informação Científica e Tecnológica em Saúde (icict)/Fiocruz, 2017, www.arca.fiocruz.br /handle/icict/34614.

11. Catalina Lopez-Quintero et al., "Probability and Predictors of Transition from First Use to Dependence on Nicotine, Alcohol, Cannabis, and Cocaine: Results of the National Epidemiologic Survey on Alcohol and Related Conditions (NESARC)," *Drug and Alcohol Dependence*, vol. 115, no. 1–2 (May 1, 2011): pp. 120–30.

12. Eliseu Labigalini et al., "Therapeutic Use of Cannabis by Crack Addicts in Brazil," *Journal of Psychoactive Drugs*, vol. 31, no. 4 (1999): pp. 451–55.

13. Gilberto Dimenstein, "Descobriram a cura do crack?," *Folha de S.Paulo*, May 24, 2010, www1.folha.uol.com.br /folha/pensata/gilbertodimenstein/739803-descobriram -a-cura-do-crack.shtml.

14. Andrea Donatti Gallassi et al., "Cannabidiol Compared to Pharmacological Treatment as Usual for Crack Use Disorder: A Feasibility, Preliminary Efficacy, Parallel, Double-Blind, Randomized Clinical Trial," *International Journal of Mental Health and Addiction* (2024), doi .org/10.1007/s11469-024-01287-z.

15. Bob Green et al., "Cannabis Use and Misuse Prevalence Among People with Psychosis," *The British Journal of Psychiatry*, vol. 187, no. 4 (2005): pp. 306–13, doi.org/10.1192 /bjp.187.4.306.

16. Anna Mané et al., "Relationship Between Cannabis and

Psychosis: Reasons for Use and Associated Clinical Variables," *Psychiatry Research*, vol. 229, no. 1–2 (September 30, 2015): pp. 70–74.

17. Berta Moreno-Küstner et al., "Prevalence of Psychotic Disorders and Its Association with Methodological Issues. A Systematic Review and Meta-Analyses," *PLoS One*, vol. 13, no. 4 (April 12, 2018): e0195687.

18. United Nations, *World Drug Report 2022*, 2022, www.unodc.org/unodc/en/data-and-analysis/world-drug-report-2022.html.

19. Antonio Waldo Zuardi et al., "A Critical Review of the Antipsychotic Effects of Cannabidiol: 30 Years of a Translational Investigation," *Current Pharmaceutical Design*, vol. 18, no. 32 (2012): pp. 5131–40.

20. Philip McGuire et al., "Cannabidiol (CBD) as an Adjunctive Therapy in Schizophrenia: A Multicenter Randomized Controlled Trial," *American Journal of Psychiatry*, vol. 175, no. 3 (March 2018): pp. 225–31.

21. R. Martin-Santos et al., "Acute Effects of a Single, Oral Dose of Delta-9-Tetrahydrocannabinol (THC) and Cannabidiol (CBD) Administration in Healthy Volunteers," *Current Pharmaceutical Design*, vol. 18, no. 32 (2012): pp. 4966–79.

22. Avshalom Caspi et al., "Moderation of the Effect of Adolescent-Onset Cannabis Use on Adult Psychosis by a Functional Polymorphism in the Catechol-O-Methyltransferase Gene: Longitudinal Evidence of a Gene X Environment Interaction," *Biological Psychiatry*, vol. 57, no. 10 (May 15, 2005): pp. 1117–27. See also Hywel J. Williams et al., "Is *COMT* a Susceptibility Gene for Schizophrenia?," *Schizophrenia Bulletin*, vol. 33, no. 3 (May 2007): pp. 635–41; and Thomas Stephanus Johannes Vaessen et al., "The Interaction Between Cannabis Use and the Val-158Met Polymorphism of the COMT Gene in Psychosis:

A Transdiagnostic Meta-Analysis," *PLoS One*, vol. 13, no. 2 (February 14, 2018): e0192658.

23. Beng-Choon Ho et al., "Cannabinoid Receptor 1 Gene Polymorphisms and Marijuana Misuse Interactions on White Matter and Cognitive Deficits in Schizophrenia," *Schizophrenia Research*, vol. 128, no. 1–3 (May 2011): pp. 66–75. See also Paula Suárez-Pinilla et al., "Brain Structural and Clinical Changes after First Episode Psychosis: Focus on Cannabinoid Receptor 1 Polymorphisms," *Psychiatry Research: Neuroimaging*, vol. 233, no. 2 (August 30, 215): pp. 112–19; and Maitane Oscoz-Irurozqui et al., "Cannabis Use and Endocannabinoid Receptor Genes: A Pilot Study on Their Interaction on Brain Activity in First-Episode Psychosis," *International Journal of Molecular Sciences*, vol. 24, no. 8 (April 19, 2023): p. 7501.

24. Ashley C. Proal et al., "A Controlled Family Study of Cannabis Users with and Without Psychosis," *Schizophrenia Research*, vol. 152, no. 1 (January 2014): pp. 283–88.

25. Joëlle A. Pasman et al., "GWAS of Lifetime Cannabis Use Reveals New Risk Loci, Genetic Overlap with Psychiatric Traits, and a Causal Effect of Schizophrenia Liability," *Nature Neuroscience*, vol. 21 (2018): pp. 1161–70.

Maturana, Marijuana, and the Green Frog

1. Humberto Maturana and Francisco Varela, *A árvore do conhecimento*, 8th ed. (São Paulo: Palas Athena, 2001).

2. Callie Barrons, "25 Inspirational Quotes About Weed," *High Times*, archived January 22, 2021, at web.archive.org/web/20210122151320/https://hightimes.com/culture/inspirational-quotes-about-weed/.

3. Barrons, "25 Inspirational Quotes About Weed."

4. Zach Fenech, "73 Weed Quotes from Influential Voices

Throughout History," *Herb*, July 25, 2024, herb.co/news /culture/weed-quotes/.

5. Yu Tse Heng et al., "Cannabis Use Does Not Increase Actual Creativity but Biases Evaluations of Creativity," *Journal of Applied Psychology*, vol. 108, no. 4 (2023): pp. 635–46.

6. Michael A. Kowal et al., "Cannabis and Creativity: Highly Potent Cannabis Impairs Divergent Thinking in Regular Cannabis Users," *Psychopharmacology*, vol. 232, no. 6 (2015): pp. 1123–34.

7. Emily M. LaFrance and Carrie Cuttler, "Inspired by Mary Jane? Mechanisms Underlying Enhanced Creativity in Cannabis Users," *Consciousness and Cognition*, vol. 56 (November 2017): pp. 68–76.

8. Gráinne Schafer et al., "Investigating the Interaction Between Schizotypy, Divergent Thinking and Cannabis Use," *Consciousness and Cognition*, vol. 21, no. 1 (2012): pp. 292–98, doi.org/10.1016/j.concog.2011.11.009.

9. Kyle S. Minor et al., "Predicting Creativity: The Role of Psychometric Schizotypy and Cannabis Use in Divergent Thinking," *Psychiatry Research*, vol. 220, no. 1–2 (December 15, 2014): pp. 205–10.

10. Russell Eisenman et al., "Undergraduate Marijuana Use as Related to Internal Sensation Novelty Seeking and Openness to Experience," *Journal of Clinical Psychology*, vol. 36, no. 4 (October 1980): pp. 1013–19.

11. Barrons, "25 Inspirational Quotes About Weed."

12. Tarcísio Alves Barreto Filho, *Cannabis medicinal para cães e gatos* (São Paulo: Manole, 2023).

13. Chris D. Verrico et al., "A Randomized, Double-Blind, Placebo-Controlled Study of Daily Cannabidiol for the Treatment of Canine Osteoarthritis Pain," *Pain*, vol. 161, no. 9 (September 2020): pp. 2191–202. See also Giorgia Della Rocca and Alessandra Di Salvo, "Hemp in

Veterinary Medicine: From Feed to Drug," *Frontiers in Veterinary Science*, vol. 7 (July 27, 2020): p. 387; Cindy H. J. Yu and H. P. Vasantha Rupasinghe, "Cannabidiol-Based Natural Health Products for Companion Animals: Recent Advances in the Management of Anxiety, Pain, and Inflammation," *Research in Veterinary Science*, vol. 140 (November 2021): pp. 38–46; and Tácio de Mendonça Lima et al., "Use of Cannabis in the Treatment of Animals: A Systematic Review of Randomized Clinical Trials," *Animal Health Research Reviews*, vol. 23, no. 1 (June 15, 2022): pp. 25–38.

Loving the Flowers Too Much

1. Roberto Catanzaro et al., "Irritable Bowel Syndrome and Lactose Intolerance: The Importance of Differential Diagnosis. A Monocentric Study," *Minerva Gastroenterology*, Turin, vol. 67, no. 1 (March 2021): pp. 72–78. See also Christian Løvold Storhaug et al., "Country, Regional, and Global Estimates for Lactose Malabsorption in Adults: A Systematic Review and Meta-Analysis," *Lancet: Gastroenterology and Hepatology*, vol. 2, no. 10 (October 2017): pp. 738–46.

2. Tamara L. Wall et al., "Biology, Genetics, and Environment: Underlying Factors Influencing Alcohol Metabolism," *Alcohol Research: Current Reviews*, vol. 38, no. 1 (2016): pp. 59–68.

3. Mimy Y. Eng et al., "ALDH2, ADH1B, and ADH1C Genotypes in Asians: A Literature Review," *Alcohol Research and Health*, vol. 30, no. 1 (2007): pp. 22–27.

4. Yasmin L. Hurd et al., "Cannabis and the Developing Brain: Insights into Its Long-Lasting Effects," *Journal of Neuroscience*, vol. 39, no. 42 (October 16, 2019): pp. 8250–58. See also Polcaro Joseph and Ivana M. Vettraino, "Cannabis in Pregnancy and Lactation. A Review," *Missouri Medicine*, vol. 117, no. 5 (September–October 2020):

pp. 400–5; and Roman Gabrhelík et al., "Cannabis Use During Pregnancy and Risk of Adverse Birth Outcomes: A Longitudinal Cohort Study," *European Addiction Research*, vol. 27, no. 2 (2021): pp. 131–41.

5. George C. Patton et al., "Cannabis Use and Mental Health in Young People: Cohort Study," *BMJ*, vol. 325, no. 7374 (November 23, 2002): pp. 1195–98. See also Michael T. Lynskey et al., "A Longitudinal Study of the Effects of Adolescent Cannabis Use on High School Completion," *Addiction*, vol. 98, no. 5 (May 2003): pp. 685–92; Andrew Lac and Jeremy W. Luk, "Testing the Amotivational Syndrome: Marijuana Use Longitudinally Predicts Lower Self-Efficacy Even After Controlling for Demographics, Personality, and Alcohol and Cigarette Use," *Prevention Science*, vol. 19, no. 2 (2018): pp. 117–26; and Aria S. Petrucci et al., "A Comprehensive Examination of the Links Between Cannabis Use and Motivation," *Substance Use & Misuse*, vol. 55, no. 7 (2020): pp. 1155–64.

6. Matthew D. Albaugh et al., "Association of Cannabis Use During Adolescence with Neurodevelopment," *JAMA Psychiatry*, vol. 78, no. 9 (2021): pp. 1031–40.

7. Daniel Feingold and Aviv Weinstein, "Cannabis and Depression," in *Cannabinoids and Neuropsychiatric Disorders*, eds. Eric Murillo-Rodriguez et al., vol. 1264, *Advances in Experimental Medicine and Biology* (Cham, Switzerland: Springer, 2021): pp. 67–80.

8. Leah Gillett et al., "Arrhythmic Effects of Cannabis in Ischemic Heart Disease," *Cannabis and Cannabinoid Research*, vol. 8, no. 5 (October 2023): pp. 867–76.

9. Helen Senderovich et al., "A Systematic Review on Cannabis Hyperemesis Syndrome and Its Management Options," *Medical Principles and Practices*, vol. 31, no. 1 (March 2022): pp. 29–38.

10. James Jett et al., "Cannabis Use, Lung Cancer, and Related

Issues," *Journal of Thoracic Oncology*, vol. 13, no. 4 (April 2018): pp. 480–87. See also Kathryn Gracie and Robert J. Hancox, "Cannabis Use Disorder and the Lungs," *Addiction*, vol. 116, no. 1 (January 2021): pp. 182–90.

11. E. B. De Sousa Fernandes Perna et al., "Subjective Aggression During Alcohol and Cannabis Intoxication Before and After Aggression Exposure," *Psychopharmacology*, vol. 233, no. 18 (2016): pp. 3331–40; and Bruna Brands et al., "Acute and Residual Effects of Smoked Cannabis: Impact on Driving Speed and Lateral Control, Heart Rate, and Self-Reported Drug Effects," *Drug and Alcohol Dependence*, vol. 205 (December 1, 2019): p. 107641.

12. Ulrich W. Preuss et al., "Cannabis Use and Car Crashes: A Review," *Frontiers in Psychiatry*, vol. 12 (May 27, 2021): p. 643315.

13. Carl Sagan and Lester Grinspoon, *Marihuana Reconsidered* (Cambridge: Harvard University Press, 1971).

14. Sonia Ortiz-Peregrina et al., "Comparison of the Effects of Alcohol and Cannabis on Visual Function and Driving Performance. Does the Visual Impairment Affect Driving?," *Drug and Alcohol Dependence*, vol. 237 (August 1, 2022): p. 109538.

15. Thomas R. Arkell et al., "Effect of Cannabidiol and Delta-9-Tetrahydrocannabinol on Driving Performance: A Randomized Clinical Trial," *JAMA*, vol. 324, no. 21 (December 1, 2020): pp. 2177–86.

16. Michael G. Lenné et al., "The Effects of Cannabis and Alcohol on Simulated Arterial Driving: Influences of Driving Experience and Task Demand," *Accident Analysis and Prevention*, vol. 42, no. 3 (May 2010): pp. 859–66. See also Tatiana Ogourtsova et al., "Cannabis Use and Driving-Related Performance in Young Recreational Users: A Within-Subject Randomized Clinical Trial," *CMAJ Open*, vol. 6, no. 4 (October 14, 2018): p. E453–62.

17. Thomas R. Arkell et al., "Cannabidiol (CBD) Content in Vaporized Cannabis Does Not Prevent Tetrahydrocannabinol (THC)-Induced Impairment of Driving and Cognition," *Psychopharmacology*, vol. 236, no. 9 (2019): pp. 2713–24.

18. Andrew Fares et al., "Combined Effect of Alcohol and Cannabis on Simulated Driving," *Psychopharmacology*, vol. 239, no. 5 (2022): pp. 1263–77.

Banning the Flowers

1. Silvia Ramos et al., *Máquina de moer gente preta: a responsabilidade da branquitude* (Rio de Janeiro: Rede de Observatórios da Segurança/cesec, 2022). See also Silvia Ramos et al., *Negro trauma: racismo e abordagem policial no Rio de Janeiro* (Rio de Janeiro: CESEC, 2022).

2. Bette Lucchese, "Dois anos após a morte de Ágatha Felix, mãe ainda aguarda julgamento de PM: 'Muita dor,'" G1, September 21, 2021, g1.globo.com/rj/rio-de-janeiro /noticia/2021/09/21/dois-anos-apos-a-morte-de-agatha -felix-mae-ainda-aguarda-julgamento-de-pm-muita-dor .ghtml.

3. Renato Filev, in discussion with the author, August 29, 2023.

4. Mano Brown, "Entrevista com Sidarta Ribeiro," *Mano a mano*, season 2, episode 5, April 21, 2022, open.spotify .com/episode/0ml5hCcOCu3LP5Pt2wkFnZ?si=43dabf 3382f8440f.

5. Cristiano Maronna, *Lei de Drogas interpretada na perspectiva da liberdade* (São Paulo: Contracorrente, 2022).

6. Part of this paragraph was first published in the magazine *CartaCapital* in April 2023.

7. Helena Martins, "Lei de drogas tem impulsionado

encarceramentonoBrasil,"*AgênciaBrasil*,lastmodifiedJune8, 2018, agenciabrasil.ebc.com.br/geral/noticia/2018-06 /lei-de-drogas-tem-impulsionado-encarceramento-no -brasil.

8.   Orlando Zaccone, *Acionistas do nada: quem são os traficantes de drogas*, vol. 2 (Rio de Janeiro: Revan, 2007).

9.   Ricardo Nemer, in discussion with the author, September 18, 2023.

10. "SBPC encaminha moção por política de drogas progressista e não proibicionista," *Sociedade Brasileira para o Progresso da Ciência*, August 10, 2018, portal.sbpcnet.org.br/noticias /sbpc-encaminha-mocao-por-politica-de-drogas-progressista -e-nao-proibicionista/.

11. Stephanie M. Zellers et al., "Recreational Cannabis Legalization Has Had Limited Effects on a Wide Range of Adult Psychiatric and Psychosocial Outcomes," *Psychological Medicine*, vol. 53, no. 14 (2023): pp. 6481–90.

Getting Old with the Flowers

1.   Nicholas Lintzeris et al., "Medicinal Cannabis in Australia, 2016: The Cannabis as Medicine Survey (CAMS-16)," *Medical Journal of Australia*, vol. 209, no. 5 (September 2018): pp. 211–16. See also Nicholas Lintzeris et al., "Medical Cannabis Use in Australia: Consumer Experiences from the Online Cannabis as Medicine Survey 2020 (CAMS-20)," *Harm Reduction Journal*, vol. 19 (July 30, 2022): p. 88; and Michelle Sexton et al., "A Cross-Sectional Survey of Medical Cannabis Users: Patterns of Use and Perceived Efficacy," *Cannabis and Cannabinoid Research*, vol. 1, no. 1 (2016): pp. 131–38.

2.   Marian S. McDonagh et al., "Cannabis-Based Products for Chronic Pain: A Systematic Review," *Annals of Internal Medicine*, vol. 175, no. 8 (2022): pp. 1143–53. See also

Roger Chou et al., "Living Systematic Review on Cannabis and Other Plant-Based Treatments for Chronic Pain: 2022 Update," *Comparative Effectiveness Review*, Rockville, no. 250 (September 2022): report no. 22-EHC042.

3. Kylie O'Brien et al., "Preliminary Findings from Project Twenty21 Australia: An Observational Study of Patients Prescribed Medicinal Cannabis for Chronic Pain, Anxiety, Posttraumatic Stress Disorder and Multiple Sclerosis," *Drug Science, Policy and Law*, vol. 9 (2023).

4. Hajar Mikaeili et al., "Molecular Basis of *FAAH-OUT*-Associated Human Pain Insensitivity," *Brain*, vol. 146, no. 9 (September 2023): pp. 3851–65.

5. Vinicius Lemos, "A pequena cidade brasileira que tinha maconha plantada até na praça principal," BBC News, September 14, 2018, www.bbc.com/portuguese /brasil-45475933.

6. Mariana Babayeva and Zvi G. Loewy, "Cannabis Pharmacogenomics: A Path to Personalized Medicine," *Current Issues in Molecular Biology*, vol. 45, no. 4 (April 17, 2023): pp. 3479–514.

7. Kifah Blal et al., "The Effect of Cannabis Plant Extracts on Head and Neck Squamous Cell Carcinoma and the Quest for Cannabis-Based Personalized Therapy," *Cancers (Basel)*, vol. 15, no. 2 (January 13, 2023): p. 497.

8. GHMedical, "About Us," ghmedical.com/mission-statement.

Dying and Being Reborn with the Flowers

1. Bridget H. Highet et al., "Tetrahydrocannabinol and Cannabidiol Use in an Outpatient Palliative Medicine Population," *American Journal of Hospice and Palliative Medicine*, vol. 37, no. 8 (2020): pp. 589–93. See also Knud Gastmeier

and Anne Gastmeier, "Niedrig dosiertes THC in der Geriatrie und Palliativmedizin" (Low-Dose THC in Geriatric and Palliative Patients), *MMW – Fortschritte der Medizin*, vol. 164, suppl. 5 (October 2022): pp. 10–14.

2. Gavin N. Petrie et al., "Endocannabinoids, Cannabinoids and the Regulation of Anxiety," *Neuropharmacology*, vol. 195 (September 1, 2021): p. 108626.

3. B. J. Miller, "An Honest Look at Marijuana and Its Place in Palliative Care," *CAPC*, August 30, 2022, www.capc.org/blog/an-honest-look-at-marijuana-and-its-place-in-palliative-care/.

4. Callie Barrons, "25 Inspirational Quotes About Weed," *High Times*, archived January 22, 2021, at web.archive.org/web/20210122151320/https://hightimes.com/culture/inspirational-quotes-about-weed/.

5. Mary Jane Gibson, "The High Times interview: Melissa Etheridge," *High Times*, archived December 8, 2024, at web.archive.org/web/20241208122540/https://hightimes.com/culture/the-high-times-interview-melissa-etheridge/.

6. Carl Sagan and Lester Grinspoon, *Marihuana Reconsidered* (Cambridge: Harvard University Press, 1971).

7. Valmir Moratelli, "Gilberto Gil: 'A maconha ajudou a minha música,'" *Quem*, July 11, 2014, revistaquem.globo.com/Entrevista/noticia/2014/07/gilberto-gil-maconha-ajudou-minha-musica.html.

8. Pacer Stacktrain, "Maya Angelou's Love of Cannabis," *Leaf Nation*, October 1, 2021, leafmagazines.com/culture/maya-angelous-love-of-cannabis/.

Epilogue

1. Carl Sagan and Lester Grinspoon, *Marihuana Reconsidered* (Cambridge: Harvard University Press, 1971).

2. Callie Barrons, "25 Inspirational Quotes About Weed," *High Times*, archived January 22, 2021, at web.archive.org /web/20210122151320/https://hightimes.com/culture /inspirational-quotes-about-weed/.

3. Gabriel Pitor, "Morre o ativista Ras Geraldinho, aos 63 anos, em Americana," *Liberal*, November 27, 2022, liberal.com.br/cidades/americana/morre-o-ativista -ras-geraldinho-aos-63-anos-em-americana-1871555/.

4. Marina Rossi, "O cientista condecorado que acabou na delegacia por causa de um líder rastafari," *El País*, February 27, 2018, brasil.elpais.com/brasil/2018/02/27/politica /1519749794_845442.html.

5. Ailton Krenak, "Roda Viva | Ailton Krenak | 19/04/2021," Roda Viva, April 19, 2021, YouTube video, 1:32:40, www .youtube.com/watch?v=BtpbCuPKTq4&t=107s.

SIDARTA RIBEIRO is a neuroscientist and biologist, postdoctoral fellow in neurophysiology at Duke University, professor at the Federal University of Rio Grande do Norte (UFRN), and member of the Centro de Estudos Estratégicos Fiocruz mental health research group. He is the author of *The Oracle of Night: The History and Science of Dreams* and *Sonho manifesto: dez exercícios urgentes de otimismo apocalíptico* (Manifest Dream: Ten Urgent Exercises in Apocalyptic Optimism). Find out more at neuro.ufrn.br/labsonhos.

DANIEL HAHN is a writer, editor, and translator with about a hundred books to his name. Previous translations include Sidarta Ribeiro's *The Oracle of Night*. He is currently translating a Guatemalan novel and writing a book about Shakespeare.